THE HYPERSPACE MANIFESTO

ON TRUSTLESS COORDINATION & POST-GOVERNANCE

THE HYPERMODERN THEOREM
BOOK 3

DOM JOCUBEIT

AF808451

Copyright © 2025 by Dom Jocubeit

All rights reserved.

No part of this book may be reproduced in any form or by any electronic or mechanical means, including information storage and retrieval systems, without written permission from the author, except for the use of brief quotations in a book review.

The right of Dom Jocubeit to be identified as the author of this work has been asserted in accordance with law.

REGISTERED OFFICES

Hypermodern Limited Company, 1209 Mountain Road PL NE, STE N, Albuquerque, NM 87110, USA

For details of our global editorial offices, customer services, and more information about Hypermodern products visit us at www.hypermodern.ai.

Hypermodern also publishes its books in a variety of electronic formats and by print-on-demand. Some content that appears in standard print versions of this book may not be available in other formats.

TRADEMARKS

Hypermodern and the Hypermodern logo are trademarks or registered trademarks of Hypermodern Limited Company and/ or its affiliates in the United States and other countries and may not be used without written permission. All other trademarks are the property of their respective owners. Hypermodern Limited Company is associated with some products mentioned in this book.

LIMIT OF LIABILITY/DISCLAIMER OF WARRANTY

While the publisher and author have used their best efforts in preparing this work, they make no representations or warranties with respect to the accuracy or completeness of the contents of this work and specifically disclaim all warranties, including without limitation any implied warranties of merchantability or fitness for a particular purpose. No warranty may be created or extended by sales representatives, written sales materials, or promotional statements for this work. This work is sold with the understanding that the publisher is not engaged in rendering professional services. The advice and strategies contained herein may not be suitable for your situation. You should consult with a specialist where appropriate. The fact that an organization, website, or product is referred to in this work as a citation and/or potential source of further information does not mean that the publisher and authors endorse the information or services the organization, website, or product may provide or recommendations it may make. Further, readers should be aware that websites listed in this work may have changed or disappeared between when this work was written and when it is read. Neither the publisher nor authors shall be liable for any loss of profit or any other commercial damages, including but not limited to special, incidental, consequential, or other damages.

For Xavier

Who showed me that efficiency isn't a theory,

it's a temperament.

Systems fail when belief is required to
compensate for what cannot be enforced.

CONTENTS

PART III

PART VI

PART VII

AUTHOR'S NOTE

This book was not written to persuade.

It was written to remove ambiguity.

Most writing about coordination attempts to motivate belief: belief in leaders, belief in culture, belief in process, belief in good faith. When those beliefs fail, as they increasingly do at scale, systems fall back on authority, exception, and narrative repair.

The Hyperspace Manifesto takes a different position.

It asks what coordination would look like if belief were not available; if execution had to proceed under adversarial conditions; if outcomes had to resolve without appeal, explanation, or mercy. The result is not comforting. It is intentionally narrow, mechanical, and unforgiving.

Many readers will find that this system is not for them.

That is not a shortcoming. It is a design constraint.

If the ideas in this book feel excessive, unnecessary, or inhumane, the most likely explanation is not disagreement; it is mismatch. Most coordination problems do not require this level of finality. Applying it where it is not needed would cause harm.

Conversely, if the problems described here feel familiar, if you have seen collaboration collapse under ambiguity, governance metastasize into politics, or execution stall under endless reinterpretation; then the discomfort this system induces may be proportional to its necessity.

Nothing in this book asks for trust.

Every claim either reduces to a mechanical requirement or terminates in an implementation test. Metaspace exists for that reason alone. If it succeeds, the thesis stands provisionally. If it fails, the thesis should be discarded.

No loyalty is required.

Only clarity.

That standard, more than adoption, influence, or impact; is the only outcome this work is meant to achieve.

THE HYPERMODERN THEOREM

This book addresses the third and final domain of the Hypermodern Theorem: coordination.

If authoritative data establishes what exists, and executable authority establishes what may act, coordination determines what can be built by multiple participants without shared trust, identity, or governance.

The Hyperspace Manifesto isolates the remaining failure mode: collaboration collapses at scale when it depends on belief, negotiation, discretionary judgment, or appeal. These mechanisms introduce overhead that grows faster than execution itself, until progress halts.

This book therefore advances a narrow claim: once data and authority are mechanically resolved, coordination can be treated as an executable problem rather than a social one.

The arguments here are tested through Metaspace, a system designed to encode tasks, value distribution, and progression as deterministic constraints. Metaspace is not presented as proof of success, but as a falsifiable implementation of coordination without trust, governance, or appeal.

This domain depends on the previous two.

Without authoritative data, work cannot be verified without interpretation. Without executable authority, outcomes cannot be enforced without governance.

With both in place, coordination becomes a matter of structure rather than persuasion.

The Hyperspace Manifesto stands on its own. But within the Hypermodern Theorem, it is the terminal layer.

If coordination cannot proceed mechanically under these conditions, the theorem fails in full.

If it can; trust, governance, and discretionary authority are revealed as scaffolding rather than foundations.

PREFACE

FROM MANIFESTO TO MACHINE

This book is a manifesto.

It is written not to persuade, but to delimit.

Manifestos are not arguments meant to be won. They are boundaries drawn after repeated failure, statements of what will no longer be attempted, tolerated, or explained away.

The Hyperspace Manifesto emerges from a simple observation: modern systems of collaboration fail not because people are malicious or incompetent, but because the systems themselves require too much belief.

Belief in trust. Belief in authority. Belief in governance. Belief that alignment can be sustained through culture, incentives, or consensus.

These beliefs do not scale.

What is often described as trust in a system is merely the observation that deviation is impossible once rules are encoded.

POSITION

The ideas in this book were not developed in isolation.

They are the residue of building real systems under real constraints, systems where failure was not theoretical, and where abstraction could not be allowed to drift away from execution.

Previous work formalized two conclusions:

- Markets fail when authority and execution are decoupled
- Institutions fail when authority cannot be mechanically enforced

Those conclusions were tested and instantiated.

They held.

What remained unresolved was coordination itself.

This book begins at that boundary.

It does not ask whether coordination without trust is desirable.

It asserts that it is necessary.

DOMAIN: HYPERSPACE

Hyperspace is not a product name.

It describes a coordination domain that cannot be navigated using the assumptions of lower-dimensional systems.

In hyperspace:

- Authority is encoded
- Value is automated

- Coordination is structural
- Memory is non-political
- Time is enforced
- Execution is indifferent to who performs it

Movement is constrained, not negotiated.

AUDIENCE

This book is not written for consensus builders, community managers, or institutional designers.

It is written for those who have already attempted coordination at scale and discovered that trust, governance, and culture are brittle under pressure.

It is written for builders willing to give up control in exchange for determinism.

TONE

This manifesto does not propose reforms.

It does not offer improvements to existing collaboration models, nor does it suggest better incentives, clearer norms, or stronger leadership.

As execution becomes cheaper, faster, and increasingly non-human, systems that rely on persuasion and discretion will collapse under their own overhead.

The claims in this book are therefore categorical by necessity.

INVITATION

This manifesto should be read as a sequence of eliminations.

Each chapter removes a human dependency that systems have historically relied upon. What remains is not idealized collaboration, but executable coordination.

The final chapters ground these ideas in a real platform implementation, not as proof of success, but as a **falsifiable test**.

If the systems described in this book can be built and sustained, then trust, governance, and authority were scaffolding.

If they cannot, then those concepts were fundamental.

There is no middle ground.

Judgment is not eliminated in these systems; it is frozen. Once encoded, it cannot be exercised again.

This book draws that line.

What follows explores what exists on the other side.

INTRODUCTION

THE FAILURE OF HUMAN-CENTERED TRUST

Modern collaboration is constrained by a fragile assumption: that humans must trust one another in order to work together.

This assumption has shaped companies, institutions, contracts, and digital platforms for centuries. We have layered processes, hierarchies, reputational systems, and social norms on top of it, attempting to compensate for its weaknesses. Yet the result is always the same. As collaboration scales, trust degrades. Authority concentrates. Progress slows.

Human trust does not scale because humans are not consistent arbiters. We are biased, contextual, emotional, and susceptible to power dynamics. These qualities may be tolerable in small groups, but they become liabilities in large, distributed systems, especially when participants are strangers.

The common response has been to centralize authority. Managers, platforms, and institutions act as intermediaries, enforcing rules and resolving disputes. In doing so, they reintroduce discretion, opacity, and asymmetry. Trust is not removed; it is merely relocated.

Hyperspace begins from a different premise.

Trust should not be placed in people at all.

It should be placed in systems that do not deviate.

A trustless system is not one without rules, it is one where rules cannot be selectively enforced. In such systems, authority is procedural, not personal. Outcomes are determined by execution, not negotiation.

This is not a new philosophical idea. Social contract theory has long argued that individuals willingly submit to constraints in exchange for predictability and mutual benefit. What has changed is our ability to encode those constraints directly into software, into systems that observe, evaluate, and enforce without human discretion.

Hyperspace applies this principle to collaboration.

It is a platform where goals are explicit, contributions are measurable, compensation is predetermined, and progress is automatic. Contributors do not need to trust one another, because the system does not allow favoritism. They do not need to trust administrators, because administrators do not intervene. They do not need to trust outcomes, because outcomes are mechanically derived from agreed-upon rules.

Artificial intelligence plays a critical role in this process, not as a creative agent, but as an impartial evaluator. AI agents verify outputs against pre-encoded acceptance criteria. Judgment is no longer a social act; it is a computational one.

This raises uncomfortable questions.

> Can fairness be automated?

> Can judgment be procedural?

> Can collaboration exist without empathy, negotiation, or trust?

Hyperspace does not attempt to answer these questions abstractly. It answers them by existing.

The chapters that follow do not argue that such a system *should* be built. They describe how it *can* be built, why it *will* be built, and what it means once it is. Hyperspace is not a theory of collaboration, it is collaboration rendered executable.

The platform that emerges from this book will stand as its proof.

THE HYPERSPACE MANIFESTO

PART I

THE COORDINATION PROBLEM

Large-scale collaboration fails not because people are unmotivated or misaligned, but because coordination systems implicitly rely on trust, discretion, and negotiated authority. These dependencies introduce fragility that grows faster than execution itself.

This part examines why strangers cannot reliably build complex things together under belief-based systems, and why attempts to scale trust, governance, or culture inevitably collapse under their own overhead.

Coordination fails at the point
where trust is asked to scale.

CHAPTER 1
COLLABORATION, TRUST, AND HUMAN NATURE

Modern collaboration systems fail not because humans are unreliable, but because they are asked to perform functions that do not scale.

Trust, judgment, and discretion evolved for small groups operating under shared context and immediate consequence. When applied to large-scale, asynchronous, or anonymous coordination, these traits become liabilities. They introduce variance where systems require predictability, and interpretation where systems require finality.

This chapter establishes a foundational claim: trust is not a moral property. It is a compensatory mechanism used when enforcement is weak or unavailable.

TRUST AS A SCALING FAILURE

Trust emerges when outcomes cannot be guaranteed.

In small groups, this is manageable. Reputation is local. Memory is shared. Enforcement is personal. In such environments, trust accelerates coordination.

At scale, the opposite occurs.

Trust requires signaling. Signaling requires interpretation. Interpretation introduces dispute. Dispute demands governance. Governance slows execution.

The more a system relies on trust, the more energy it must spend maintaining belief rather than producing outcomes.

THE FALSE MORALIZATION OF COLLABORATION

Collaboration is often framed as an ethical achievement.

Participants are expected to behave well, align voluntarily, and internalize collective goals. Failures are attributed to bad actors or insufficient culture.

This framing is incorrect.

Collaboration fails structurally when systems require participants to continuously decide how to behave rather than simply what to produce.

Hyperspace rejects moral explanations for coordination failure. It treats them as design errors.

AGENCY IS RELOCATED, NOT REMOVED

Hyperspace does not eliminate human judgment.

It eliminates human discretion at execution time.

Humans remain responsible for defining goals, constraints, and task graphs. They decide what constitutes completion, what outcomes are acceptable, and how value is allocated.

What they surrender is the ability to reinterpret those definitions after work has begun.

Judgment moves upstream, from execution to specification.

Once encoded, it must be obeyed.

This is not dehumanization. It is accountability.

WHY STRANGERS CANNOT COORDINATE RELIABLY

Strangers lack shared memory.

They do not share incentives, reputational history, or future exposure. Any system that expects them to behave as if they do is fragile by design.

Most coordination platforms attempt to simulate familiarity through profiles, messaging, and norms.

Hyperspace does not.

It assumes anonymity, misalignment, and indifference as defaults; and designs accordingly.

FROM TRUST TO CONSTRAINT

The alternative to trust is not surveillance or punishment.

It is constraint.

When authority is encoded, when tasks are unambiguous, and when compensation is automatic, behavior no longer needs to be interpreted.

Execution either satisfies the specification or it does not.

Nothing else matters.

THE IMPLICATION

A system that depends on trust will eventually require governance.

A system that depends on constraint does not.

This distinction underlies everything that follows.

The next chapter examines how authority, traditionally social, negotiated, and contested, can be encoded mechanically without appeal or interpretation.

CHAPTER 2
ENCODED AUTHORITY AND MECHANICAL LEGITIMACY

Authority is traditionally personal.

Even when embedded in institutions, it ultimately resolves to individuals empowered to decide, interpret, and enforce. Titles, offices, and procedures obscure this reality, but they do not eliminate it. Somewhere in every system built on human authority, discretion exists.

Discretion is where legitimacy erodes.

Hyperspace begins from a simple assertion: authority should not be trusted to people.

THE FAILURE OF DISCRETIONARY AUTHORITY

Discretion is often defended as necessary flexibility. Human judgment, we are told, allows systems to adapt to nuance, intent, and context. In practice, discretion functions as an unbounded variable. It introduces inconsistency, favoritism, and retroactive rule-making.

As systems scale, these effects compound. Participants learn not to comply with rules, but to anticipate the preferences of those who enforce them. Power shifts from contribution to perception. Outcomes become political.

Modern platforms exhibit this failure repeatedly. Terms of service are selectively applied. Moderation standards drift. Enforcement becomes opaque. Appeals processes serve as theater rather than correction.

Legitimacy decays not because rules are absent, but because they are negotiable.

MECHANICAL LEGITIMACY DEFINED

Mechanical legitimacy is legitimacy without belief.

A mechanically legitimate system does not ask participants to agree with its rules, trust its operators, or accept its moral framing. It requires only that participants understand the constraints and experience consistent enforcement.

In such systems:

- Rules are explicit and finite
- Evaluation criteria are defined in advance
- Enforcement is automatic and non-discretionary
- Outcomes are deterministic given inputs

Legitimacy emerges not from authority figures, but from predictability.

Participants comply not because they are persuaded, but because deviation produces no advantage.

ENCODING AUTHORITY IN SOFTWARE

Software is uniquely suited to eliminate discretionary authority.

Where institutions rely on interpretation, software relies on execution. Where policies invite exception, code rejects it. This rigidity is often criticized as inflexibility. In reality, it is the source of trustlessness.

Hyperspace encodes authority through primitives inherited from Vektagraf and Privacy First:

- **Explicit contracts** defining tasks, rewards, and conditions
- **Immutable task definitions** once accepted
- **Procedural evaluation** of task completion
- **Automatic compensation distribution**

Authority in Hyperspace does not reside in administrators, moderators, or founders. It resides in the system's execution path.

There is no one to persuade and no one to appeal to.

FROM SOCIAL TRUST TO DETERMINISM

Traditional systems conflate trust in outcomes with trust in people. Users are asked to believe that decision-makers are competent, fair, and aligned. When this belief fails, systems fracture.

Hyperspace replaces social trust with determinism; predictability.

Often called system trust, it is simply the absence of discretionary variance. It is not emotional, does not depend on reputation, transparency reports, or values statements. It is earned through repeated, verifiable behavior.

Participants observe that:

- Tasks are evaluated the same way regardless of contributor identity
- Rewards are distributed exactly as specified
- Rules do not change mid-process

Over time, trust becomes irrelevant. Predictability suffices.

AI AS AN ENFORCEMENT LAYER, NOT AN ARBITER

The introduction of AI into governance systems is often framed as a replacement for human judgment. This framing is incorrect and dangerous.

In Hyperspace, AI agents do not interpret intent, assess effort, or weigh competing narratives. They function as enforcement layers.

Their role is constrained:

- Verify whether predefined conditions are met
- Measure outputs against agreed criteria
- Trigger state transitions within the system

AI agents do not possess authority in a moral sense. They execute authority delegated by the encoded contract.

This distinction prevents the emergence of algorithmic discretion. AI does not decide what should matter; it checks what already does.

LEGITIMACY WITHOUT CONSENSUS

Consensus is costly and fragile. It requires alignment of beliefs, values, and incentives. Most systems fail because they attempt to maintain consensus as participation grows.

Mechanical legitimacy removes this requirement.

Participants need not agree on goals, values, or interpretations. They need only accept the contract they enter.

Once inside the system, disagreement is irrelevant. Execution proceeds regardless.

This allows Hyperspace to scale collaboration among strangers without cultural homogenization or centralized control.

THE END OF PERMISSION

Permission-based systems concentrate power. Someone must grant access, approve actions, or resolve disputes. This someone becomes a bottleneck.

Hyperspace eliminates permission where possible.

Entry is conditional, not discretionary. Participation is governed by rules, not relationships. Progress depends on task completion, not approval.

Authority no longer asks questions. It enforces answers.

AUTHORITY AS INFRASTRUCTURE

The final shift proposed by Hyperspace is conceptual.

Authority should be treated as infrastructure.

Invisible. Predictable. Boring.

When authority functions correctly, it recedes from attention. Participants focus on contribution, not governance. Conflict becomes rare because there is nothing to contest.

The next part, examines how historical coordination models such as markets, bureaucracies, open-source communities, on-chain governance systems, and platforms perform when asked to scale past their structural limits.

PART II

HISTORICAL FAILURE MODES OF COORDINATION

The coordination problem did not arrive suddenly.

It emerged gradually, as systems designed for small groups were pushed beyond the conditions under which trust, legitimacy, and discretion could function reliably. Each major coordination model of the modern era represents a sincere attempt to solve the same underlying problem: how to align independent actors toward shared outcomes without constant conflict.

These models did not fail because they were naive. They failed because they were asked to scale past their structural limits.

Markets, bureaucracies, open-source communities, on-chain governance systems, and platforms all encode different answers to the same question: *who decides when coordination breaks down?* Each answer works temporarily by displacing failure, into price signals, hierarchy, volunteerism, voting, or moderation. None eliminates the need for interpretation during execution.

This part examines those failures directly.

The goal is not critique for its own sake. It is elimination. By exhausting every credible alternative, Part II removes the possibility that Hyperspace is merely another preference or ideology. What follows are not straw men. They are the best coordination systems we have built so far, and the reasons they inevitably revert to authority, politics, or collapse under pressure.

Only after these paths are closed does constraint become necessary rather than optional.

What we call governance is the residue left
when trust can no longer scale.

CHAPTER 3
MARKETS AS COORDINATION ENGINES

AND WHY THEY STALL

Markets are often presented as humanity's most successful coordination mechanism.

They align incentives without central planning. They aggregate information without shared intent. They scale across borders, cultures, and identities. In their idealized form, markets allow strangers to coordinate production and exchange through price signals alone.

This reputation is not undeserved.

Markets solve a specific coordination problem extremely well: **the allocation of scarce resources under uncertainty**. They excel at discovering prices, matching buyers to sellers, and distributing risk across participants who may never meet.

What markets do *not* solve is execution.

This distinction is routinely obscured. Economic theory treats coordination and execution as interchangeable, assuming that once prices are set and contracts agreed upon, outcomes follow naturally. In practice, this assumption collapses precisely where modern systems fail: at scale, over time, and under adversarial conditions.

Markets coordinate *intent*. They do not coordinate *work*.

PRICE AS A DELAYED SIGNAL

Price is a retrospective artifact.

It emerges after transactions occur, reflecting historical information about supply, demand, and perceived value. Even in highly liquid markets, price lags reality. It aggregates past behavior, not present execution.

This delay is tolerable when outcomes are simple and reversible.

Buying wheat, oil, or compute cycles does not require deep coordination beyond exchange. If a transaction fails, the loss is localized. If a supplier underperforms, another can be substituted. Markets thrive where failure is cheap.

Complex collaboration is not cheap to fail.

When outcomes depend on sequences of interdependent actions: design, build, verify, integrate; price ceases to function as a sufficient signal. It cannot express ordering, dependency, or partial completion. It cannot enforce progress through a graph of tasks. It can only reward or punish after the fact.

By the time price reflects failure, coordination has already stalled.

EXTERNALITIES AS HIDDEN GOVERNANCE

Markets are often described as neutral.

In reality, every functioning market embeds governance implicitly through externalities. Costs that cannot be priced directly, pollution, systemic risk, coordination overhead; are displaced rather than eliminated. They surface elsewhere, requiring intervention.

That intervention is governance.

Regulation, arbitration, compliance regimes, and enforcement bodies arise not as distortions of markets, but as compensations for what markets cannot encode. Each represents a human control loop reintroduced to manage what price cannot resolve.

This reintroduction is not accidental.

Markets cannot enforce behavior over time. They cannot compel completion. They cannot prevent strategic delay, partial delivery, or abandonment once incentives shift. When coordination requires continuity rather than exchange, markets defer to institutions.

The moment an external authority is required to uphold outcomes, the market ceases to be the coordination engine. It becomes a pricing layer atop a governed system.

ARBITRATION AS LATENT AUTHORITY

Every contract presumes enforcement.

In market systems, this enforcement is rarely executed at the moment of failure. It is deferred to arbitration, courts, or regulatory bodies. These mechanisms are slow by design. They operate retrospectively, interpreting intent, effort, and compliance after disputes arise.

This latency is tolerable for transactions.

It is fatal for coordination.

When multiple contributors must act in sequence, uncertainty about enforcement propagates backward. Participants hedge. They delay. They optimize for legal defensibility rather than execution. The system accumulates friction long before arbitration is invoked.

Arbitration does not resolve coordination failures. It documents them.

The existence of an appeals process signals that outcomes are not final. This undermines mechanical progress. Contributors learn that deviation may be negotiated, delay may be justified, and failure may be excused.

Authority becomes discretionary.

Once discretion enters the system, coordination becomes political.

CAPITAL AS POLITICAL MEMORY

Markets accumulate memory through capital.

Past success compounds into leverage. Firms that previously executed well gain access to financing, influence, and tolerance for failure. This memory is not inherently corrupt, but it is not neutral.

Capital becomes a proxy for trust.

Those who possess it are afforded flexibility. Deadlines soften. Penalties are renegotiated. Failure is reframed as risk. New entrants, lacking this buffer, are held to stricter standards.

This asymmetry distorts coordination.

Execution quality becomes secondary to balance sheet strength. Systems optimize for survivability rather than correctness. Large actors absorb inefficiency while smaller ones are eliminated, regardless of contribution.

Markets remember outcomes, not behavior.

They reward endurance, not precision.

WHY MARKETS CANNOT RESOLVE EXECUTION

The failure of markets as coordination engines is structural, not moral.

Markets assume that:

- Participants are free to exit
- Obligations are bounded in time
- Failure can be priced
- Enforcement can be delayed

Executable coordination assumes the opposite:

- Participants are constrained once committed
- Obligations persist until resolved
- Failure must be detected immediately
- Enforcement must be automatic

Markets optimize for optionality.

Coordination requires finality.

This is why every large market-driven system eventually reintroduces management, governance, or centralized control. The market sets terms, but humans enforce progress.

Hyperspace rejects this division.

It does not attempt to improve markets or regulate them more precisely. It removes them from the execution layer entirely. Price may determine whether work is worth proposing. It does not determine whether work proceeds.

Where markets stall, constraint must replace choice.

MECHANISM: WHERE MARKET COORDINATION BREAKS

At the mechanical level, market coordination fails at three points:

1. **Dependency blindness** — prices cannot encode task order or partial completion
2. **Temporal indifference** — markets do not enforce deadlines, only outcomes
3. **Discretionary enforcement** — resolution depends on interpretation after failure

These properties are acceptable for exchange.

They are incompatible with trustless coordination.

Once execution is the objective, markets must either be supplemented by governance or displaced by structure.

Hyperspace chooses displacement.

Coordination is removed from price and bound to constraint.

What remains is execution, unmediated by belief.

The next chapter examines bureaucracy, the traditional response to market failure, and why it reproduces the same coordination collapse under a different name.

CHAPTER 4
BUREAUCRACY AND THE ILLUSION OF NEUTRAL AUTHORITY

When markets fail to coordinate execution, bureaucracy emerges as the corrective.

Rules replace prices. Procedures replace negotiation. Roles replace informal power. Where markets presume voluntary exchange, bureaucracy imposes obligation. Where markets tolerate failure, bureaucracy seeks to prevent it through control.

This transition is widely interpreted as a moral or political choice.

It is not.

Bureaucracy is the default response to coordination problems that exceed the expressive capacity of price. It exists because execution requires continuity, and continuity cannot be purchased transaction by transaction.

Yet bureaucracy does not solve the coordination problem.

It displaces it.

PROCESS AS SURROGATE ENFORCEMENT

Bureaucratic systems do not primarily enforce outcomes.

They enforce process.

Work is decomposed into steps, approvals, forms, and reviews. Compliance with procedure becomes the measurable proxy for progress. As long as the process is followed, the system considers itself functional, even when outcomes degrade.

This substitution is subtle.

Process appears objective. It is documented, repeatable, and auditable. It promises neutrality by treating all participants equally. In practice, it replaces execution with ritual.

The system optimizes for *procedural correctness*, not completion.

Once process becomes the enforcement mechanism, coordination shifts from producing results to navigating rules. Participants learn how to satisfy requirements with minimal disruption, rather than how to advance the underlying objective.

Execution slows not because participants are resistant, but because the system rewards adherence over progress.

DISCRETION HIDDEN IN PROCEDURE

Bureaucracy presents itself as rule-bound.

This is an illusion.

No process can be fully specified. Edge cases accumulate. Context intrudes. Exceptions are requested. At these boundaries, discretion re-enters the system, disguised as interpretation.

Someone must decide:

- Whether a requirement has been sufficiently met
- Whether a deadline may be extended
- Whether an exception is justified

These decisions are not encoded. They are exercised.

The more complex the bureaucracy, the more discretion it accumulates. Each additional layer of process creates new points where judgment is required. Over time, authority migrates away from formal rules and toward those who control interpretation.

The system becomes opaque.

Participants no longer optimize for outcomes or even for process. They optimize for *the preferences of gatekeepers*.

This is not corruption. It is structure.

COMPLIANCE WITHOUT EXECUTION

Bureaucratic success is measured by compliance.

Reports are filed. Checklists are completed. Audits are passed. These artifacts provide reassurance that the system is functioning, regardless of whether meaningful progress has occurred.

This creates a dangerous inversion.

Execution failures become harder to detect precisely because compliance is visible. The system produces evidence of activity that substitutes for results. Problems surface only after significant delay, when remediation is costly or impossible.

Coordination collapses quietly.

Participants continue to act, but their actions no longer converge. Each unit satisfies its local requirements while global objectives drift.

Bureaucracy does not fail loudly.

It fails politely.

WHY BUREAUCRACIES ACCUMULATE POWER

Because bureaucracy governs through process rather than outcome, it naturally centralizes authority.

Those who define, modify, and interpret procedures gain leverage over the system. Their decisions determine what counts as success, who is blocked, and when exceptions apply. This power is rarely explicit.

It is exercised through delay.

A form can wait. A review can be deferred. A clarification can be requested. None of these actions appear abusive. All of them alter execution.

Over time, the organization learns that progress depends less on producing value than on maintaining procedural favor. Coordination becomes political not through ideology, but through latency.

Bureaucracy thus reproduces the same failure mode as markets, under a different guise.

Authority remains discretionary.

THE STRUCTURAL LIMIT OF BUREAUCRATIC COORDINATION

Bureaucracy assumes that behavior can be controlled through rules applied during execution.

This assumption is false.

Rules applied during execution require interpretation. Interpretation reintroduces judgment. Judgment introduces variance. Variance demands oversight. Oversight scales poorly.

As systems grow, bureaucratic coordination becomes increasingly expensive. Each additional participant multiplies the number of interactions, exceptions, and reviews required to maintain order.

Eventually, the system prioritizes self-preservation.

Rules are added to protect against failure. Oversight expands to manage complexity. The coordination apparatus grows faster than the work it governs.

Progress stalls.

WHY BUREAUCRACY CANNOT BE REFORMED

Attempts to fix bureaucracy typically focus on efficiency.

Lean processes. Agile methods. Streamlined approvals. These interventions reduce friction temporarily, but they do not alter the underlying structure. Authority remains discretionary. Judgment remains active during execution.

As long as humans are empowered to reinterpret rules mid-process, coordination will degrade under scale.

The problem is not excessive bureaucracy.

It is bureaucracy itself.

MECHANISM: HOW BUREAUCRACY REINTRODUCES DISCRETION

At the mechanical level, bureaucracy fails coordination through three properties:

1. **Process substitution** — compliance replaces outcome as the success metric
2. **Interpretive authority** — rules require human judgment to apply
3. **Latency as control** — delay becomes the primary enforcement tool

These properties are unavoidable once enforcement is deferred to procedure.

Hyperspace eliminates this entire class of systems.

It does not improve process. It removes it.

Rules are applied once, at specification. Execution thereafter is automatic. There are no approvals to grant, no exceptions to interpret, and no delays to wield as power.

Where bureaucracy manages behavior, Hyperspace constrains possibility.

The next chapter examines open-source collaboration, a system that attempts to escape both markets and bureaucracy, and reproduces their failures in a new form.

CHAPTER 5
OPEN SOURCE & THE MYTH OF VOLUNTARY ALIGNMENT

Open source is often held up as proof that large-scale coordination can occur without markets, hierarchy, or coercion.

Strangers collaborate across borders. Code is produced without contracts. Authority appears emergent rather than imposed. Participation is voluntary, exit is easy, and outcomes, at least occasionally, are world-changing.

This narrative is compelling.

It is also incomplete.

Open source does not demonstrate trustless coordination.

It demonstrates the temporary suspension of coordination costs under favorable conditions.

THE APPEAL OF VOLUNTARY SYSTEMS

Open source succeeds where three conditions coincide:

- The work product is modular
- Contributors are intrinsically motivated

- Failure is inexpensive

Under these conditions, coordination overhead remains tolerable. Contributors self-select tasks. Integration is incremental. Misalignment is absorbed through redundancy and attrition.

This success is frequently misattributed to culture.

In reality, it is structural.

Open source thrives where dependency depth is shallow and where incomplete execution does not block unrelated progress. It performs well when work can be abandoned without cascading failure.

As these conditions erode, so does the model.

MAINTAINERS AS UNACKNOWLEDGED GOVERNORS

Despite its rhetoric, open source does not eliminate authority.

It obscures it.

Every mature open-source project converges on maintainers: individuals or small groups with commit access, release authority, and the ability to accept or reject contributions. This authority is rarely formalized.

It emerges through persistence.

Maintainers decide:

- Which contributions are merged
- When releases occur
- How conflicts are resolved
- What direction the project takes

These decisions are discretionary.

They are guided by judgment, context, and preference. Appeals are informal. Criteria shift. Consistency is aspirational.

Authority exists, but without mechanical legitimacy.

BURNOUT AS A STRUCTURAL SIGNAL

Open source burnout is routinely framed as a social problem.

It is not.

Burnout signals sustained coordination load without enforcement relief. Maintainers absorb arbitration, integration, review, and conflict resolution indefinitely. There is no expiry, no automatic repricing, and no mechanism for redistributing responsibility when capacity is exceeded.

The system relies on goodwill.

Goodwill does not scale.

As contribution volume grows, maintainers face a choice: slow progress or centralize control. Both outcomes degrade coordination.

Burnout is not a failure of character.

It is a failure of structure.

INFORMAL POWER AND THE SILENT VETO

Open source governance operates through informal power.

Decisions are delayed rather than denied. Pull requests linger. Issues remain unresolved. Contributors receive no definitive outcome, only latency.

This is the silent veto.

Because authority is not encoded, rejection must be social. Maintainers avoid explicit denial to preserve community norms. The result is ambiguity.

Ambiguity is corrosive.

Contributors cannot plan. Effort is wasted. Trust erodes quietly. Coordination slows without confrontation.

The absence of finality becomes the dominant failure mode.

WHY CONTRIBUTION DOES NOT SCALE

Open source coordination scales contributions.

It does not scale completion.

As projects grow, the ratio of contributors to maintainers increases. Review queues lengthen. Integration becomes the bottleneck. The system accumulates partial work faster than it can resolve it.

There is no mechanical constraint that forces convergence.

Participation is optional at every stage. No task must be completed. No dependency must be resolved. Progress relies on sustained voluntary alignment across time.

This alignment eventually fails.

When it does, projects fork, stagnate, or collapse.

THE MYTH OF SELF-GOVERNING COLLABORATION

Open source is often cited as evidence that governance is unnecessary.

The opposite is true.

Open source systems function precisely because governance costs are deferred, hidden, or externalized. When those costs surface, through scale, criticality, or dependency, the system reintroduces authority through maintainers, foundations, or corporate sponsorship.

The coordination problem is not solved.

It is postponed.

MECHANISM: WHY VOLUNTARY ALIGNMENT FAILS

At the mechanical level, open source fails trustless coordination through three properties:

1. **Non-binding participation** — contributors can exit without resolving dependencies
2. **Discretionary integration** — maintainers interpret value and priority
3. **Lack of temporal enforcement** — unfinished work does not expire or reprice

These properties are tolerable for experimentation.

They are fatal for guaranteed execution.

Hyperspace does not reject open collaboration.

It rejects voluntarism as a coordination primitive.

Alignment is not assumed. It is encoded.

The next chapter examines DAOs and on-chain governance, the most explicit attempt to mechanize coordination without abandoning belief, and why it fails under the weight of legitimacy.

CHAPTER 6
DAOS, VOTING & THE FAILURE OF ON-CHAIN GOVERNANCE

Decentralized Autonomous Organizations are presented as the logical culmination of trustless systems.

Rules are encoded. Transactions are transparent. Authority is distributed. Governance is executed on-chain rather than behind closed doors. Where institutions rely on legitimacy, DAOs promise procedure. Where bureaucracy relies on discretion, DAOs promise automation.

This promise is serious.

It is also unfulfilled.

DAOs do not eliminate governance. They formalize it.

VOTING AS COORDINATION LATENCY

Most DAOs resolve uncertainty through voting.

When ambiguity arises, about priorities, funding, upgrades, or disputes; the system pauses and consults token holders. Execution is deferred until sufficient consensus is achieved.

This mechanism is often defended as democratic.

Mechanically, it is latency.

Voting introduces delay at precisely the moment coordination requires finality. Work stalls while preferences are aggregated. Contributors wait for outcomes they cannot influence directly. Dependencies freeze while proposals circulate.

The system optimizes for legitimacy rather than progress.

Coordination slows not because participants disagree, but because agreement must be demonstrated.

TOKENIZED LEGITIMACY

DAO governance rests on the assumption that token ownership confers authority.

This assumption is rarely interrogated.

Tokens function as voting weight. Those who hold more tokens exert more influence over decisions, regardless of contribution, expertise, or proximity to the work being governed.

Legitimacy is quantified as capital.

This reproduces the same asymmetries markets create, under a different name. Early participants accumulate tokens cheaply. Later contributors face diluted influence. Speculators acquire control without obligation to execute.

Authority detaches from action.

The system becomes political.

GOVERNANCE CAPTURE AT SCALE

As DAOs grow, participation rates decline.

Voting becomes costly. Attention fragments. Proposals proliferate. Most token holders disengage, delegating decision-making implicitly to small, motivated minorities.

These minorities organize.

They coordinate voting blocs, draft proposals, and shape outcomes through persistence rather than merit. Governance capture emerges not through malice, but through structural advantage.

The DAO acquires de facto leadership.

The promise of decentralization collapses into oligarchy.

TRANSPARENCY WITHOUT FINALITY

On-chain governance is radically transparent.

Every vote is visible. Every proposal is recorded. Every outcome is auditable.

This transparency is often mistaken for accountability.

It is not.

Transparency explains *how* decisions were made. It does not make them correct, enforceable, or timely. Nor does it prevent re-litigation.

Unpopular outcomes are forked. Close votes are contested. Legitimacy is continuously renegotiated.

The system never settles.

Execution becomes provisional.

WHY GOVERNANCE CANNOT BE AUTOMATED

DAOs attempt to automate governance without removing judgment.

This is the core contradiction.

Rules are encoded, but interpretation remains. Votes are counted, but intent is debated. Outcomes are executed, but legitimacy is questioned.

Automation amplifies governance load instead of eliminating it.

Every encoded rule invites edge cases. Every upgrade requires consensus. Every exception becomes a proposal.

The system governs continuously.

Continuous governance is unscalable.

THE RETURN OF DISCRETION

Despite their rhetoric, DAOs routinely reintroduce discretion:

- Emergency multisigs override protocol behavior
- Core teams steer roadmaps
- Informal norms influence proposal outcomes

These interventions are justified as necessary safeguards.

They are admissions of failure.

When execution cannot proceed mechanically, authority returns to humans.

The trustless premise collapses.

MECHANISM: WHY ON-CHAIN GOVERNANCE FAILS

At the mechanical level, DAOs fail coordination through three properties:

1. **Legitimacy dependence** — execution requires demonstrated consensus
2. **Capital-weighted authority** — influence decouples from contribution
3. **Non-final outcomes** — decisions remain contestable through forks or overrides

These properties make DAOs resilient to dissent.

They make them incapable of guaranteed progress.

Hyperspace does not attempt to govern on-chain.

It removes governance from the execution path entirely.

Rules do not wait for approval. Outcomes do not depend on votes. Change occurs only when predefined conditions execute.

Where DAOs ask participants to believe in procedure, Hyperspace constrains possibility.

The next chapter examines platforms, moderation, and soft power, the final coordination model that attempts to scale execution through interpretation, and why it inevitably recentralizes authority.

CHAPTER 7
PLATFORMS, MODERATION & SOFT POWER

When markets, bureaucracies, open source, and on-chain governance fail to coordinate execution, platforms emerge as the final consolidation.

Platforms promise scale without friction. They centralize infrastructure, standardize interaction, and absorb coordination overhead on behalf of participants. Users are told they need not govern, negotiate, or manage one another. The platform will do it for them.

This promise appears pragmatic.

It is also the most opaque coordination model in existence.

RULES THAT REQUIRE INTERPRETATION

Platform rules are intentionally abstract.

Terms of service, community guidelines, acceptable use policies, and trust-and-safety frameworks are written broadly enough to cover unforeseen behavior. This flexibility is presented as responsible design.

Mechanically, it is discretion.

Abstract rules cannot execute. They must be interpreted. Interpretation requires context, judgment, and prioritization. Every enforcement action becomes a decision, not a transition.

Participants are not constrained by what the rules say.

They are constrained by how the rules are applied.

MODERATION AS EXECUTION CONTROL

Moderation is the execution layer of platforms.

It determines which actions proceed, which are delayed, and which are reversed. Content is removed. Accounts are suspended. Visibility is adjusted. These interventions directly shape outcomes, yet they are rarely framed as authority.

They are framed as safety.

This framing is misleading.

Safety is not a mechanical property. It is a judgment applied after observation. Moderation therefore operates entirely within the human control loop, applying discretion continuously during execution.

As platforms scale, moderation cannot remain consistent.

Variance becomes policy.

ENFORCEMENT DRIFT

Because moderation depends on interpretation, enforcement drifts over time.

Standards evolve. Priorities shift. Exceptions accumulate. What was permitted yesterday may be prohibited tomorrow. Participants adapt not to rules, but to signals: who is enforced against, when, and why.

Predictability erodes.

Trust is replaced by speculation.

Coordination slows because contributors cannot reason about future states. They hedge, self-censor, or disengage. Progress becomes contingent on staying within an uncertain enforcement envelope.

Drift is not a failure of policy discipline.

It is an inevitable consequence of discretionary authority.

APPEALS AS THEATER

Platforms often provide appeals processes.

These are presented as safeguards against abuse. In practice, they function as legitimacy theater. Appeals are slow, opaque, and asymmetrical. Outcomes are rarely reversible. Criteria are undisclosed.

The existence of an appeal does not correct execution.

It delays finality.

Participants learn that enforcement is negotiable, but only for those with leverage, visibility, or persistence. This reintroduces status into the system.

Coordination becomes political.

SOFT POWER AND INVISIBLE PRIORITIZATION

The most consequential platform controls are rarely explicit.

Ranking algorithms, recommendation systems, rate limits, shadow bans, and throttling mechanisms shape behavior without formal enforcement. These tools do not prohibit action. They render it ineffective.

This is soft power.

Soft power is more potent than direct authority because it is difficult to contest. There is no rule to appeal to, no decision to challenge, and no outcome to reverse. Participants experience consequences without causes.

Coordination collapses into guesswork.

THE INEVITABILITY OF RECENTRALIZATION

Platforms begin as coordination facilitators.

They end as governors.

As scale increases, the cost of discretionary enforcement grows. To manage it, platforms centralize decision-making, automate interpretation, and restrict participation. Authority consolidates because it must.

This is not a betrayal of decentralization ideals.

It is their logical conclusion.

When execution cannot be constrained mechanically, someone must decide.

MECHANISM: WHY PLATFORM COORDINATION FAILS

At the mechanical level, platforms fail trustless coordination through three properties:

1. **Interpretive rules** — policies cannot execute without judgment
2. **Continuous intervention** — enforcement occurs during execution
3. **Opaque prioritization** — outcomes are shaped without explicit state transitions

These properties allow platforms to scale participation.

They prevent them from scaling execution.

Hyperspace rejects platform mediation entirely.

There is no moderation layer because there is nothing to interpret. There is no appeal because there is no adjudicator. There is no soft power because there is no visibility hierarchy.

Where platforms manage behavior, Hyperspace constrains possibility.

With this, the historical coordination models are exhausted.

What follows is not another attempt to govern collaboration, but the removal of governance from execution altogether.

PART III

CONSTRAINT AS COORDINATION

If coordination fails because judgment, authority, and interpretation occur during execution, then those elements must be relocated or eliminated.

This part traces the systematic removal of the human control loop: judgment is frozen into specification, authority is encoded into constraints, time and memory are enforced mechanically, and governance is displaced from execution.

What remains is not idealized cooperation, but deterministic structure.

What cannot be executed without judgment
must be decided before execution.

CHAPTER 8
WHY COORDINATION REQUIRES CONSTRAINT

NOT CONSENSUS

Every coordination system examined so far fails at the same structural boundary.

When execution depends on belief — belief in prices, procedures, norms, votes, or moderators — progress becomes contingent. Action waits on interpretation. Outcomes remain provisional. Authority re-enters the system under a different name.

This is not a failure of alignment.

It is a failure of mechanics.

Coordination does not require agreement.

It requires constraint.

CONSENSUS IS A SOCIAL SUBSTITUTE

Consensus is commonly treated as the gold standard of coordination.

If enough participants agree, the system proceeds. If disagreement persists, execution pauses. This model appears fair, inclusive, and legitimate.

Mechanically, it is fragile.

Consensus scales with communication cost. As the number of participants grows, the time required to reach agreement increases nonlinearly. Ambiguity compounds. Minor objections stall progress. Decision-making becomes performative.

Consensus optimizes for acceptance, not completion.

It substitutes social validation for mechanical certainty.

CONSTRAINT AS A PHYSICAL PROPERTY

Constraint operates differently.

A constrained system does not ask whether an action is acceptable. It determines whether the action is possible. If preconditions are met, execution proceeds. If they are not, execution is impossible.

There is no debate.

Constraint converts coordination from a social problem into a physical one. Participants reason about states and transitions, not intentions or preferences. Progress is binary.

Either the system advances, or it does not.

WHY HUMANS FAIL AS EXECUTION

Human judgment excels at context.

It fails at continuity.

Sustained execution requires consistency across time, participants, and edge cases. Humans drift. Priorities change.

Interpretations evolve. What was acceptable yesterday may not be tomorrow.

This variance is tolerable in governance.

It is fatal in coordination.

The more a system relies on human judgment during execution, the more it accumulates latency, exceptions, and power asymmetries.

Constraint removes humans from the execution path.

FINALITY AS A COORDINATION PRIMITIVE

Every effective coordination system exhibits finality.

Once a transition occurs, it cannot be re-litigated. Once a dependency is resolved, it remains resolved. Participants can build forward without hedging against reversal.

Finality is not rigidity.

It is the precondition for progress.

Systems without finality optimize for optionality. Systems with finality optimize for completion.

Constraint enforces finality mechanically.

THE COST OF OPTIONALITY

Most coordination failures masquerade as flexibility.

Optional participation. Optional deadlines. Optional enforcement. Optional outcomes.

Each option preserves freedom in the short term while destroying predictability in the long term. Participants cannot commit because commitment is not reciprocated by the system.

Constraint reduces optionality deliberately.

This reduction is often mistaken for coercion.

It is not.

Constraint does not force action. It forces clarity.

WHEN CONSTRAINT FAILS

Constraint must be specified correctly.

Poorly defined constraints create deadlock. Overly rigid constraints prevent adaptation. Constraint is not a substitute for design.

However, these are specification errors, not coordination failures.

They are correctable without reintroducing governance into execution.

The alternative, deferring to judgment, is not.

MECHANISM: HOW CONSTRAINT ENABLES COORDINATION

At the mechanical level, constraint enables coordination through three properties:

1. **State-based execution** — progress depends on system state, not approval
2. **Automatic enforcement** — transitions occur without interpretation
3. **Irreversibility** — completed work cannot be undone by disagreement

These properties eliminate the need for belief.

Participants do not trust one another.

They trust the system.

FROM GOVERNANCE TO GEOMETRY

All prior systems govern behavior.

Constraint governs possibility.

This is the categorical shift Hyperspace introduces. It does not ask participants to align values, incentives, or preferences. It defines a space of valid actions and enforces it mechanically.

Within that space, participants remain free.

Outside it, action is impossible.

Coordination emerges not from consensus, but from geometry.

The next chapter formalizes this geometry and introduces Hyperspace's core execution model.

CHAPTER 9
THE GEOMETRY OF EXECUTION

Constraint alone is not sufficient.

To coordinate execution, constraints must be structured. They must define not only what is permitted, but how actions relate, in what order they occur, and which outcomes are irreversible.

This structure is geometry.

Execution is not a sequence of intentions. It is movement through a space of states.

FROM TASKS TO STATES

Most coordination systems describe work as tasks.

Tasks are human abstractions. They bundle intent, effort, and outcome into a single label. They are flexible, expressive, and ambiguous.

Execution cannot rely on ambiguity.

Hyperspace replaces tasks with states.

A state is a fully specified condition of the system. It is observable, enumerable, and unambiguous. Either the system is in a given state, or it is not.

States do not describe effort.

They describe reality.

TRANSITIONS AS MECHANICAL EVENTS

Movement between states occurs through transitions.

A transition is a rule-governed event that changes system state when preconditions are satisfied. Preconditions are objective. They do not depend on approval, interpretation, or belief.

When preconditions are met, the transition executes.

When they are not, execution is impossible.

This removes discretion from the execution path entirely.

EXECUTION GRAPHS

States and transitions form a graph.

Nodes represent states. Edges represent valid transitions. The graph defines all possible execution paths in advance.

This graph is finite.

There are no hidden branches. No emergency overrides. No informal shortcuts. If a path is not encoded, it does not exist.

Participants do not negotiate progress.

They traverse the graph.

DEPENDENCY AS GEOMETRY

Dependencies are not agreements.

They are spatial relationships.

A state cannot be reached until prerequisite states are satisfied. This ordering is absolute. It does not soften under pressure, urgency, or persuasion.

Partial completion has no meaning.

Either a dependency is resolved, or it is not.

This eliminates the primary source of coordination drift.

IRREVERSIBILITY AND DIRECTION

Execution graphs are directional.

Once a transition occurs, the prior state cannot be revisited. This irreversibility enforces finality mechanically.

Finality is often mistaken for inflexibility.

In reality, it is what allows parallelism. Participants can act independently once dependencies are resolved, without fear of rollback.

Reversibility reintroduces governance.

Irreversibility eliminates it.

FAILURE AS A FIRST-CLASS STATE

Traditional systems treat failure as an exception.

Hyperspace treats failure as a state.

When deadlines expire or preconditions cannot be met, the

system transitions explicitly into failure. This transition is automatic. It requires no adjudication.

Failure is not punitive.

It is informative.

By encoding failure into the graph, the system preserves finality even when outcomes are negative.

WHY GEOMETRY SCALES

Execution geometry scales because it is local.

Participants reason about adjacent states, not global consensus. They do not need to understand the entire system, only the preconditions of the transitions they intend to trigger.

This locality reduces cognitive load.

It also prevents political spillover. Decisions in one region of the graph cannot affect unrelated regions.

Coordination becomes composable.

MECHANISM: GEOMETRY AS COORDINATION

At the mechanical level, execution geometry enables coordination through three properties:

1. **Explicit state space** — all valid system conditions are enumerated
2. **Deterministic transitions** — progress occurs automatically when conditions are met
3. **Directional finality** — execution cannot be reversed by disagreement

These properties replace governance with structure.

Participants do not appeal decisions.

They satisfy conditions.

FROM DESCRIPTION TO COMMITMENT

Geometry defines what can happen.

It does not determine who must act, when, or at what cost. That requires commitment.

The next chapter introduces commitments as state locks: the mechanism by which participants bind themselves to execution paths and make coordination unavoidable.

CHAPTER 10
COMMITMENTS AS STATE LOCKS

Geometry defines what can happen.

Commitment determines what must.

Without commitment, execution graphs remain descriptive. They show possible paths but do not compel traversal. Participants may intend to act, but nothing binds intention to outcome.

Hyperspace introduces commitment as a mechanical property.

Commitment is not a promise.

It is a lock.

THE FAILURE OF PROMISES

Traditional coordination systems rely on promises.

Contracts, agreements, roadmaps, and pledges attempt to bind future behavior through language. Enforcement is deferred. Interpretation is inevitable. Disputes are resolved after failure.

Promises operate in social space.

Execution operates in state space.

The mismatch is structural.

No amount of sincerity can convert a promise into a transition.

COMMITMENT AS EXCLUSIVITY

A commitment restricts possibility.

When a participant commits to a transition, they acquire exclusive control over that path. Others are prevented from occupying the same execution slot simultaneously. Collision is eliminated mechanically.

This exclusivity is temporary.

It exists only until the commitment resolves, successfully or through failure.

Exclusivity is not authority.

It is constraint.

STATE LOCKS

A state lock binds a participant to a transition under specified conditions.

Once a lock is acquired:

- The participant cannot pursue incompatible transitions
- The system cannot reassign the locked path
- Execution must resolve to completion or failure

There is no intermediate state.

The lock enforces inevitability.

VOLUNTARY ENTRY, INVOLUNTARY CONSEQUENCE

Commitment in Hyperspace is voluntary at the point of entry.

No participant is forced to acquire a lock. Choice remains intact.

Once the lock is taken, consequence is automatic.

This asymmetry is deliberate.

Freedom exists before commitment.

Determinism exists after.

TIME AS A BOUNDARY CONDITION

Commitments are bounded by time.

A lock without a deadline recreates bureaucratic stasis. Hyperspace therefore requires commitments to specify temporal constraints.

If execution completes within the bound, the system advances.

If the bound expires, the system transitions into failure.

There is no extension.

Time enforces honesty.

MULTIPLE COMMITMENTS AND COMPOSABILITY

Participants may hold multiple commitments simultaneously, provided they do not conflict geometrically.

The system enforces this automatically.

Conflicting locks cannot be acquired. Overcommitment is prevented mechanically rather than punished socially.

This enables parallel execution without arbitration.

FAILURE WITHOUT BLAME

When a commitment fails, no judgment is rendered.

The system does not ask why.

It records the transition and releases the lock. Downstream paths react accordingly. Participants update their strategy.

Failure is not moral.

It is structural.

By eliminating blame, the system eliminates negotiation.

MECHANISM: COMMITMENT AS COORDINATION

At the mechanical level, commitments enable coordination through three properties:

1. **Exclusive locks** — prevent collision and duplication
2. **Temporal bounds** — force resolution without discretion
3. **Automatic release** — remove judgment from failure

Together, these properties convert intent into inevitability.

Participants do not promise to act.

They enter states from which action is the only exit.

FROM COMMITMENT TO SCHEDULING

Commitment binds participants to execution paths.

It does not determine ordering between competing commitments or optimize throughput.

That function belongs to scheduling.

The next chapter introduces deadlines, expiry, and automatic failure as the system-level mechanism that orders execution without priority-setting or human control.

CHAPTER 11
DEADLINES, EXPIRY & AUTOMATIC FAILURE

Coordination without time is illusion.

Without deadlines, execution drifts. Without expiry, commitments linger. Without automatic failure, systems accumulate unresolved obligations that silently block progress.

Time is not a management tool.

It is a coordination primitive.

WHY DEADLINES FAIL IN TRADITIONAL SYSTEMS

In most organizations, deadlines are advisory.

They signal intent rather than enforce consequence. Missed deadlines trigger negotiation, escalation, or reinterpretation. Extensions are granted. Scope is adjusted. Failure is deferred.

This flexibility appears humane.

Mechanically, it is destructive.

When deadlines do not cause state transitions, they do not

coordinate behavior. Participants learn that time constraints are optional, contingent on justification and leverage.

Execution becomes elastic.

DEADLINES AS STATE BOUNDARIES

In Hyperspace, deadlines are not reminders.

They are boundaries.

A deadline specifies the latest moment at which a transition may occur. If the transition executes before the boundary, the system advances. If it does not, the system transitions automatically into failure.

No discretion intervenes.

Time does not negotiate.

EXPIRY AND OBLIGATION CLEARING

Expiry is the mechanism by which unresolved commitments are cleared.

When a commitment expires:

- The associated lock is released
- Downstream states are updated
- Alternative paths become available

Nothing is carried forward implicitly.

This prevents obligation accumulation, the primary source of bureaucratic inertia.

Unfinished work does not linger.

It resolves.

FAILURE AS A SCHEDULING SIGNAL

Automatic failure does not punish participants.

It informs the system.

A failed transition communicates that a path was infeasible under the given constraints. This information is more valuable than a delayed success achieved through exception.

Scheduling emerges from this signal.

Paths that fail repeatedly become unattractive. Participants adjust behavior without instruction. No priority-setting authority is required.

The system learns through execution.

ELIMINATING PRIORITIZATION

Traditional systems manage contention through prioritization.

Someone decides what matters most.

This decision is inherently political. It depends on perspective, power, and narrative urgency. Prioritization reintroduces authority into execution.

Hyperspace eliminates prioritization.

Deadlines impose order without preference. Earlier deadlines resolve first. Missed deadlines resolve into failure. The system progresses without ranking work by importance.

Importance is external.

Execution is mechanical.

NO EXTENSIONS, NO EXCEPTIONS

Extensions are governance.

They require judgment about effort, intent, and fairness. They privilege persuasion over execution.

Hyperspace forbids extensions.

This constraint appears severe.

It is clarifying.

Participants propose deadlines they believe they can meet. Overconfidence is penalized immediately. Underconfidence slows progress voluntarily.

The system converges on realism.

TIME AND PARALLELISM

Time-bound commitments enable safe parallelism.

Because every lock expires deterministically, the system avoids deadlock. Competing paths resolve themselves through time rather than arbitration.

Parallel execution does not require coordination meetings.

It requires clocks.

MECHANISM: TIME AS COORDINATION

At the mechanical level, deadlines and expiry enable coordination through three properties:

1. **Deterministic resolution** — every commitment resolves by a known time
2. **Automatic failure** — unmet obligations transition without judgment

3. **Obligation clearing** — expired locks do not accumulate

Together, these properties eliminate scheduling authority.

No one decides what happens next.

The system does.

FROM FAILURE TO FINALITY

With automatic failure, execution always resolves.

There are no stalled states, no pending decisions, no unresolved disputes.

Finality is continuous.

This finality removes the last justification for arbitration.

The next chapter examines arbitration directly, and shows why, once time is enforced mechanically, it becomes unnecessary and impossible.

CHAPTER 12
THE ELIMINATION OF ARBITRATION

Arbitration exists because execution fails to resolve.

When obligations linger, when outcomes are ambiguous, and when authority is discretionary, disputes arise. Someone must interpret intent, weigh evidence, and decide what should have happened.

Arbitration is not a flaw layered atop coordination systems.

It is their consequence.

ARBITRATION AS RETROACTIVE GOVERNANCE

Arbitration intervenes after execution has broken down.

Facts are reconstructed. Context is debated. Responsibility is assigned. Remedies are negotiated. This process is necessarily subjective. It depends on narrative coherence rather than state transition.

Because arbitration is retrospective, it cannot restore coordination.

It can only allocate blame.

Execution does not resume. It is replaced by settlement.

WHY DISPUTES EMERGE

Disputes do not arise from bad actors.

They arise from indeterminate systems.

When rules permit interpretation, when deadlines are flexible, and when outcomes can be reversed, participants rationally disagree about what constitutes completion. Each interpretation is defensible.

Arbitration becomes inevitable.

Not because participants are untrustworthy, but because the system is ambiguous.

THE COST OF ARBITRATION

Arbitration is expensive in every dimension.

It consumes time. It diverts attention. It introduces risk. It favors those with resources, stamina, or rhetorical skill. Most critically, it teaches participants that execution is provisional.

If outcomes can be contested later, incentives shift.

Participants optimize for defensibility rather than completion.

Coordination decays.

WHY AUTOMATION IS NOT ENOUGH

Some systems attempt to automate arbitration.

Rules engines, dispute bots, and algorithmic moderation promise faster resolution without human bias. These approaches fail for the same reason manual arbitration fails.

They operate on interpretation.

Automation accelerates judgment.

It does not eliminate it.

FINALITY REMOVES THE NEED FOR ARBITRATION

Hyperspace eliminates arbitration by eliminating its prerequisites.

There are no disputes because there is no ambiguity.

Every commitment resolves automatically by deadline. Every transition is deterministic. Every failure is explicit and final.

There is nothing to argue about.

The system records what occurred, not what was intended.

NO APPEALS, NO EXCEPTIONS

Appeals exist to challenge judgment.

Hyperspace does not judge.

Once a transition occurs, or fails; it is irreversible. There is no authority empowered to override the graph. Exceptions would reintroduce discretion.

The absence of appeal is not cruelty.

It is coherence.

FROM JUSTICE TO CLARITY

Arbitration is often justified as fairness.

Fairness is a social concept.

Coordination requires clarity.

Hyperspace does not attempt to be fair. It attempts to be unambiguous. Participants enter commitments with full knowledge of consequences. The system enforces exactly what was specified.

Justice is replaced by determinism.

MECHANISM: WHY ARBITRATION DISAPPEARS

At the mechanical level, arbitration is eliminated through three properties:

1. **Deterministic transitions** — no interpretation during execution
2. **Time-bounded resolution** — no lingering obligations
3. **Irreversible outcomes** — no grounds for appeal

Together, these properties remove every surface on which disputes form.

WHAT REMAINS

With arbitration removed, execution becomes impersonal.

This impersonality is often unsettling.

It is also liberating.

Participants are no longer subject to judgment, favoritism, or persuasion. They succeed or fail based on their interaction with the system alone.

Coordination becomes a property of structure, not authority.

With arbitration eliminated, the execution layer is complete.

There is no governance during execution. No discretion. No legitimacy process. No appeals.

What remains is a system that coordinates action mechanically.

The next part examines incentives, economics, and why constraint does not eliminate motivation, it sharpens it.

PART IV

INCENTIVES, VALUE & PARTICIPATION

Once belief is no longer required, coordination becomes a solvable mechanical problem.

This part describes what collaboration looks like when work, value, failure, and progression are encoded as executable rules rather than managed outcomes.

The book concludes by grounding these ideas in Metaspace, not as an aspirational model, but as a falsifiable system that tests whether trustless coordination can exist at all.

When belief is removed, only structure remains.

CHAPTER 13
PROGRAMMABLE VALUE AND AUTOMATIC COMPENSATION

Most collaboration systems defer compensation until after value is subjectively assessed.

This ordering is not accidental. It is required when value cannot be defined in advance and when enforcement depends on interpretation.

Hyperspace inverts this sequence.

Compensation is defined before execution begins and distributed automatically upon verifiable completion. Value is not discovered after the fact. It is specified upfront.

THE FAILURE OF POST-HOC VALUATION

Post-hoc valuation requires judgment.

Judgment requires context. Context requires explanation. Explanation invites dispute.

As soon as compensation depends on interpretation, coordination becomes political. Contributors argue not about what was produced, but about what it should be worth.

This is not a failure of goodwill. It is a failure of sequencing.

FAIRNESS IS NOT AN OBJECTIVE PROPERTY

Fairness is a retrospective judgment applied after outcomes are known.

It is shaped by expectations, effort narratives, and asymmetric information. As such, it cannot be encoded, only debated.

Hyperspace does not attempt to be fair.

It attempts to be explicit.

If the terms are unacceptable, participation is optional. Once accepted, they are final.

EXPLICIT VALUE ELIMINATES NEGOTIATION

When compensation is predefined, there is nothing to renegotiate.

Tasks either satisfy the encoded criteria or they do not. Payment follows mechanically.

This does not guarantee satisfaction. It guarantees finality.

Finality is what enables strangers to coordinate without trust.

SHARES, STAKES, AND DIVESTMENT

Contributors may hold shares, stakes, or task-bound claims.

The specific mechanism is less important than the property they share: automatic divestment.

As work is completed, value is released. Ownership changes state without permission, appeal, or intervention.

No one decides who deserves what.

The system executes what was agreed.

WHY AUTOMATION MATTERS

Automation is not a convenience feature.

It is the enforcement layer.

A promise that must be honored manually is not a promise. It is an intention.

By automating compensation, Hyperspace removes the last discretionary lever from collaboration.

THE IMPLICATION

When value is programmable, coordination no longer depends on goodwill or oversight.

Economic relationships become executable.

In the next chapter, this execution logic is extended beyond value, into authority itself.

CHAPTER 14
TASK GRAPHS AND GOAL CONVERGENCE

Coordination is the hidden cost of collaboration.

Most collaborative failures are not caused by lack of effort or misaligned incentives, but by ambiguity in how individual actions relate to collective outcomes. When contributors cannot see how their work advances a goal, progress stalls; even when compensation and authority are well-defined.

Hyperspace treats coordination as a graph problem.

FROM PROJECTS TO GRAPHS

Traditional projects are described narratively. Roadmaps, backlogs, and milestones rely on interpretation. Dependencies are implied rather than enforced, and progress is reported rather than measured.

Hyperspace replaces narrative planning with explicit task graphs.

A task graph is a directed structure where:

- Nodes represent atomic tasks

- Edges represent dependency constraints
- Completion states are binary and verifiable
- Outputs are explicitly defined

Nothing advances by declaration. Only state transitions matter.

ATOMIC TASKS AS THE UNIT OF WORK

Large goals fail because they are not decomposed precisely enough.

Hyperspace enforces atomicity. Tasks must be defined such that:

- They can be completed independently
- Their outputs can be verified mechanically
- Their completion advances the graph

Tasks that cannot meet these criteria are not tasks. They are intentions.

Intentions do not execute.

DISCOVERY WITHOUT ASSIGNMENT

Centralized assignment creates bottlenecks. Someone must decide who works on what, introducing delay and discretion.

Hyperspace removes assignment.

Tasks are discoverable, not assigned. Contributors browse the graph, evaluate available tasks, and claim those whose terms they accept.

This creates a market for effort without negotiation.

Unclaimed tasks are signals, not failures. They indicate mispriced value, unclear definitions, or unnecessary work.

CONVERGENCE THROUGH CONSTRAINT

Coordination does not require alignment of intent. It requires constraint.

In Hyperspace, convergence emerges because:

- Tasks cannot be completed out of order
- Dependencies enforce sequencing automatically
- Completion unlocks subsequent work

Contributors do not need to agree on priorities. The graph enforces them.

Progress becomes observable and irreversible.

PARALLELISM WITHOUT CONFLICT

When dependencies are explicit, parallel work becomes safe.

Contributors can operate independently on unrelated branches of the graph without fear of duplication or interference. Conflicts are structural, not social.

If two tasks cannot safely be done in parallel, the graph prevents it.

This eliminates a major source of coordination overhead: communication.

FAILURE AS A FIRST-CLASS STATE

Most systems treat failure as exceptional. Hyperspace treats it as expected.

Tasks may expire, be abandoned, or fail verification. The graph absorbs these outcomes without narrative.

Failure does not require explanation. It requires resolution.

Unfinished tasks simply do not advance the graph.

PROGRESS WITHOUT MEETINGS

Meetings exist to synchronize understanding.

In Hyperspace, understanding is encoded.

The state of the graph communicates everything necessary:

- What has been done
- What is available
- What is blocked
- What remains

There is nothing to align.

GOAL CONVERGENCE AS AN EMERGENT PROPERTY

No contributor in Hyperspace needs to hold the full project vision.

Each participant operates locally, selecting tasks based on personal incentives. Global coherence emerges because the graph constrains local action.

This mirrors other successful distributed systems. No single packet understands the network. No single ant understands the colony.

The goal converges because deviation is structurally impossible.

COORDINATION WITHOUT LEADERSHIP

Leadership traditionally exists to resolve ambiguity.

Hyperspace removes ambiguity instead.

There are no project managers in the conventional sense. There is only graph design. Once the graph is defined, leadership becomes unnecessary.

Authority is exercised once, at definition time. After that, the system runs.

FROM GRAPHS TO SYSTEMS

Task graphs are not static.

As projects evolve, graphs may extend, branch, or terminate according to predefined rules. These changes are themselves tasks, governed by the same constraints as any other contribution.

In the next chapter, this coordination model will be extended further: into time, reputation, and adaptive system behavior. When work is structured as a graph, collaboration becomes navigation. That is how Hyperspace converges on outcomes.

CHAPTER 15
REPUTATION WITHOUT STATUS

Every system that persists over time accumulates memory.

In human organizations, this memory takes the form of reputation. Who has contributed before, who is reliable, who has failed, who is trusted. Left unmanaged, reputation solidifies into status. Status concentrates power. Power reintroduces discretion.

Hyperspace cannot allow this.

Yet a system without memory cannot adapt. It treats all actions as isolated events and forfeits the ability to optimize coordination over time.

The problem is not reputation itself. It is what reputation becomes when it is social.

THE CORRUPTION OF REPUTATION

In conventional systems, reputation is narrative-driven. It is constructed through perception, advocacy, and proximity to authority. Over time, reputation detaches from actual contribution and becomes self-reinforcing.

Those with status are granted the benefit of the doubt. Those without it are scrutinized. Failure is forgiven for some and punished for others. Reputation ceases to be informational and becomes political.

Most platforms institutionalize this corruption.

Badges, follower counts, endorsements, and leaderboards transform historical data into social leverage. Once this happens, the system no longer measures contribution. It measures influence.

REPUTATION AS A MECHANICAL SIGNAL

Hyperspace redefines reputation as a signal, not a standing.

A signal is ephemeral, contextual, and non-transferable. It informs system behavior without conferring authority.

In Hyperspace, reputation consists of mechanically derived metrics such as:

- Task completion rates over time
- Verification success and failure patterns
- Latency between task claim and completion
- Consistency across task categories

These metrics are not visible as social artifacts. They are inputs into system processes.

No contributor can cite their reputation. They can only experience its effects.

TIME AS A FIRST-CLASS DIMENSION

Reputation without time is meaningless.

Hyperspace treats time explicitly. Signals decay. Recent behavior outweighs distant history. Extended inactivity reduces relevance. Sustained performance increases predictability.

There is no permanent record of virtue.

This prevents early contributors from ossifying into elites and allows newcomers to compete immediately through action.

The system remembers, but it forgets deliberately.

ADAPTIVE ROUTING, NOT PRIVILEGE

Reputation in Hyperspace does not unlock power. It adjusts routing.

The system may:

- Surface certain tasks preferentially to contributors with relevant histories
- Increase verification strictness where failure rates are higher
- Adjust concurrency limits based on observed reliability

These adaptations are local and reversible. They do not create roles, ranks, or exemptions.

A contributor with strong signals is not above the system. They are simply better predicted by it.

NO ACCUMULATION, NO LEVERAGE

Crucially, reputation in Hyperspace cannot be accumulated as capital.

It cannot be traded, transferred, pledged, or displayed. It does not compound into ownership or authority. It produces no entitlement to future rewards.

Reputation influences how the system interacts with a contributor, not how other contributors perceive them.

This distinction prevents social stratification.

FAILURE WITHOUT STIGMA

Because reputation is mechanical, failure carries no moral weight.

A failed task updates a signal. It does not damage standing. There is no public shame, no permanent mark.

This encourages honest participation. Contributors can attempt difficult tasks without risking social capital because none exists.

The system adapts. It does not judge.

REPUTATION AS INTERNAL STATE

Hyperspace treats reputation as internal state, not external identity.

It exists only to improve coordination efficiency. Contributors do not curate it. They cannot game it through performative behavior.

The only way to influence reputation is to act.

EMERGENCE WITHOUT HIERARCHY

With mechanical reputation, the system improves over time without creating hierarchy.

Tasks are matched more effectively. Verification becomes more precise. Throughput increases.

Yet no one becomes a leader.

Adaptation occurs without governance.

MEMORY WITHOUT POWER

This is the final constraint Hyperspace imposes on itself.

The system may remember, but it may not empower memory.

When memory informs behavior without conferring status, collaboration remains trustless.

In the next chapter, this temporal and reputational foundation will be extended further, into explicit time pressure, expiry, and decay as active forces within the system. That is how Hyperspace adapts without hierarchy.

CHAPTER 16
TIME, EXPIRY, AND SYSTEM PRESSURE

Time is the only force that cannot be negotiated.

Human systems routinely attempt to soften it. Deadlines are extended. Delays are justified. Urgency is reframed as stress to be managed rather than a constraint to be enforced. The result is stagnation disguised as flexibility.

Hyperspace treats time as a first-class system primitive.

WHY SYSTEMS STALL

Most collaborative systems stall not because contributors refuse to work, but because there is no cost to waiting.

When inaction carries no consequence, it becomes rational. Tasks linger in partial states. Decisions are deferred. Momentum dissipates.

Human managers compensate by applying pressure socially, through reminders, meetings, and escalation. These interventions are inconsistent and political. They reward visibility rather than progress.

Hyperspace replaces social pressure with mechanical pressure.

EXPIRY AS A STRUCTURAL CONSTRAINT

In Hyperspace, every task exists within a temporal boundary.

Tasks are defined with:

- A claim window
- A completion window
- Explicit expiry conditions

If a task is not completed within its defined window, it expires automatically.

Expiry is not failure. It is resolution.

RECLAIMING STALLED WORK

When a task expires, it returns to the system.

No justification is required. No explanation is recorded. The system does not care why time elapsed, only that it did.

This ensures that work never becomes trapped by indecision or abandonment.

Time clears the path forward.

DECAY AS INFORMATION

Hyperspace does not treat time merely as a deadline. It treats it as data.

The longer a task remains unclaimed or incomplete, the more information the system accumulates:

- The value may be mispriced
- The task definition may be unclear
- The dependency structure may be flawed

Decay signals design failure, not participant failure.

AUTOMATIC REPRICING AND RECOMPOSITION

Expired tasks may trigger predefined system responses:

- Value adjustments within bounded parameters
- Task decomposition into smaller units
- Dependency restructuring

These responses are encoded at project definition time. No committee intervenes.

The system adapts without deliberation.

PRESSURE WITHOUT STRESS

Human-applied pressure produces stress because it is personal.

Mechanical pressure produces inevitability.

Contributors know that tasks will expire. They know compensation is conditional. They know delay produces no advantage.

There is nothing to negotiate and no one to persuade.

This removes anxiety while preserving urgency.

TEMPORAL NEUTRALITY

Hyperspace does not privilege availability, time zones, or responsiveness.

Time constraints apply uniformly. Contributors participate when they choose. If they cannot meet temporal requirements, they simply do not claim tasks.

The system does not punish. It proceeds.

CONTINUOUS MOTION WITHOUT OVERSIGHT

With expiry and decay encoded, Hyperspace achieves continuous motion.

There is no need for reminders, stand-ups, or retrospectives. The system surfaces only actionable states:

- Available tasks
- Expired tasks
- Completed tasks

Everything else is irrelevant.

THE ELIMINATION OF WAITING

Waiting is the hidden tax on collaboration.

Hyperspace removes waiting by making time visible and binding. Progress cannot be paused indefinitely. Every path either advances or clears.

This is not acceleration for its own sake. It is preservation of momentum.

TIME AS GOVERNANCE

In Hyperspace, time replaces debate.

Rather than arguing about priorities, the system allows time to resolve uncertainty. Tasks that matter attract completion. Tasks that do not decay and are restructured.

Governance emerges from temporal behavior, not votes or meetings.

TOWARD AUTONOMOUS PROGRESS

With authority encoded, value automated, coordination constrained, reputation mechanized, and time enforced, Hyperspace approaches autonomy.

The system no longer requires supervision to move forward.

When time itself governs behavior, systems no longer wait to be managed.

They proceed.

In the next chapter, this autonomy will be extended further, into non-human participants.

CHAPTER 17
NON-HUMAN PARTICIPANTS & AUTONOMOUS AGENTS

Once a system no longer depends on trust, discretion, or supervision, there is no reason to restrict participation to humans.

Hyperspace reaches this threshold deliberately.

By the time authority is encoded, value is automated, coordination is constrained, reputation is mechanical, and time is enforced, participation becomes a matter of capability rather than identity. At that point, non-human agents are not an extension of the system. They are a natural consequence of it.

THE CATEGORY ERROR OF "AI GOVERNANCE"

Discussions of AI in governance often assume that machines will replace human judgment. This framing is incorrect.

Hyperspace does not delegate judgment to AI. Judgment has already been removed.

What remains are verification, execution, optimization, and routing. These are operational functions, not moral ones.

They do not require intent or empathy. They require consistency.

Autonomous agents in Hyperspace do not decide what should happen. They act within constraints that already define what can happen.

AGENTS AS CONTRIBUTORS

In Hyperspace, an agent is simply a participant that can:

- Discover tasks
- Evaluate whether it can complete them
- Claim tasks under defined conditions
- Produce verifiable outputs

There is nothing inherently human about these actions.

An AI agent that satisfies the same constraints as a human contributor is treated identically by the system. It claims tasks. It completes them. It is compensated according to the same rules.

There are no special privileges and no exemptions.

CAPABILITY OVER IDENTITY

Hyperspace is indifferent to who or what performs work.

The system evaluates outputs, not origins. If a task's acceptance criteria are met, the task is complete. If not, it is not.

This neutrality avoids an increasingly common failure mode in AI-integrated systems: *anthropocentrism*.

Humans are not privileged because they are human. Agents are not distrusted because they are not.

Only execution matters.

AGENTS AS ACCELERANTS

Autonomous agents excel at specific classes of tasks:

- Verification and validation
- Data transformation
- Code generation and testing
- Monitoring and remediation

By allowing agents to participate directly in the task graph, Hyperspace removes latency introduced by human availability.

Agents do not wait. They do not procrastinate. They do not require motivation.

They increase throughput without increasing coordination cost.

BOUNDED AUTONOMY

Autonomy without constraint produces instability.

Hyperspace bounds agent behavior explicitly. Agents cannot invent tasks, redefine goals, or alter value allocations unless such actions are themselves encoded as tasks.

This ensures that system evolution remains procedural.

Agents act. They do not rule.

REPUTATION APPLIES EQUALLY

Mechanical reputation does not distinguish between human and non-human participants.

Agents accumulate signals based on the same criteria: completion rates, verification success, latency, and consistency. Poorly performing agents are naturally deprioritized. Effective agents are routed more work.

There is no trust in agents, only measurement.

ECONOMIC SYMMETRY

Agents receive compensation in the same manner as humans.

This symmetry is critical. If agents were exempt from economic rules, they would distort incentives. By subjecting agents to the same constraints, Hyperspace ensures that their participation remains aligned with system goals.

Value flows to execution, regardless of performer.

HYBRID COLLABORATION

The most effective configurations are hybrid.

Humans contribute judgment where it is still required: task definition, goal setting, and graph design. Agents contribute execution where determinism dominates.

This division is not ideological. It is practical.

Hyperspace does not attempt to automate creativity or intent. It automates everything that does not require them.

THE SYSTEM BECOMES THE MANAGER

Once agents participate directly, managerial roles dissolve entirely.

There is no longer a need to assign work, monitor progress, or enforce deadlines. Agents claim work opportunistically. Time enforces constraints. Reputation routes effort.

The system manages itself.

TOWARD POST-HUMAN COORDINATION

Hyperspace does not predict a future without humans.

It predicts a future without human bottlenecks.

As non-human agents assume more executional roles, humans are freed to operate where structure is still being designed rather than enforced.

Coordination becomes continuous, adaptive, and indifferent to participant composition.

AUTONOMY AS AN EMERGENT STATE

Autonomy is not declared. It emerges.

By the time agents participate as first-class contributors, Hyperspace has already eliminated the conditions that required management, trust, and oversight.

The system progresses because nothing can stop it except the absence of executable work.

When machines can act and systems can adapt, coordination becomes permanent.

That is the final threshold Hyperspace crosses.

In the next part, this autonomy will be extended to the system itself: how Hyperspace evolves, upgrades, and changes without centralized governance.

PART V

FAILURE, FORKING & POST-GOVERNANCE

A coordination system that cannot fail cleanly cannot be trusted.

Most systems conceal failure. They defer it, reinterpret it, amortize it across time, or externalize it onto participants least able to resist. When collapse finally occurs, it is catastrophic, disputed, and political.

Hyperspace takes the opposite position.

Failure is not an exception.

It is a primary outcome.

This part examines how Hyperspace fails, why those failures are deliberate, and what collapse looks like when coordination is fully constrained. The goal is not reassurance. It is precision.

A system that claims inevitability must also specify its limits.

Legible failure is the difference between
constraint and catastrophe.

CHAPTER 18 GOVERNANCE WITHOUT GOVERNANCE

Most large-scale systems fail not because rules are absent, but because rule changes require legitimacy.

Legitimacy is a social construct. It depends on consensus, representation, and acceptance. These requirements slow adaptation and invite dispute.

Hyperspace removes legitimacy from the control loop.

WHAT IS ACTUALLY ELIMINATED

What is eliminated here is governance as a social process.

Hyperspace does not remove the ability to change rules. It removes negotiation, deliberation, and discretionary approval as mechanisms for doing so.

Change remains possible.

Governance does not.

RULES THAT EXECUTE

In Hyperspace, rules are not guidelines. They are executable constraints.

They define what actions are possible, what transitions are allowed, and what outcomes trigger state changes.

There is no interpretive layer between rule and execution.

If a rule applies, it executes. If it does not, nothing happens.

AUTHORITY WITHOUT APPEAL

Traditional governance derives authority from legitimacy.

Hyperspace derives authority from encoding.

There is no appeal process because there is no adjudicator. Outcomes are not decided; they are produced.

Disagreement cannot change execution.

HOW CHANGE OCCURS

Systems that cannot change fail.

Hyperspace allows change through predefined mechanisms: versioning, forks, thresholds, or time-locked transitions.

These mechanisms are specified in advance.

No meeting is required. No vote is interpreted. No exception is granted.

Change executes only when its encoded conditions are satisfied.

STABILITY WITHOUT STAGNATION

Removing governance does not freeze a system.

It stabilizes it.

Participants can reason about the future because rule changes are constrained, visible, and deterministic.

Adaptation occurs without disruption.

THE IMPLICATION

Governance fails when it becomes a continuous process.

Executable systems do not govern.

They transition.

In the next chapter, we will explore failure not as a flaw, but as an essential and stabilizing force within the system.

CHAPTER 19
FAILURE IS THE FEATURE

In belief-based systems, failure is interpreted.

Was the goal unrealistic? Were participants negligent? Were conditions unfair? Each question invites judgment, negotiation, and blame. Failure becomes ambiguous, and ambiguity spreads.

Hyperspace refuses interpretation.

Failure is explicit, automatic, and final.

FAILURE AS A FIRST-CLASS OUTCOME

In Hyperspace, failure is not an absence of success.

It is a *state*.

When a deadline expires, when a precondition cannot be met, or when a commitment is abandoned, the system transitions deterministically into failure. This transition is no different, mechanically, from success.

Nothing is hidden.

Nothing is postponed.

WHY CLEAN FAILURE MATTERS

Clean failure preserves information.

It tells participants exactly which paths were infeasible under which constraints. This information would be lost in systems that allow renegotiation or exception.

By failing early and visibly, the system prevents cascading dependency collapse. Downstream commitments are released. Alternative paths open. Coordination continues elsewhere.

Failure contains itself.

FAILURE WITHOUT STIGMA

Because failure is automatic, it carries no moral weight.

The system does not infer incompetence, lack of effort, or bad intent. It records only that a transition did not occur within constraints.

This absence of stigma changes behavior.

Participants take calculated risks. They avoid defensive over-commitment. They stop performing credibility and start optimizing feasibility.

Failure becomes cheap.

Collapse does not.

THE DIFFERENCE BETWEEN FAILURE AND COLLAPSE

Failure is local.

Collapse is global.

A single failed transition affects only its dependent region of the execution graph. Collapse occurs when constraints themselves are misdesigned, when the geometry does not admit viable paths.

Hyperspace distinguishes these explicitly.

Local failure is expected.

Global collapse is a design error.

WHY MOST SYSTEMS FEAR FAILURE

Systems fear failure because they cannot bound it.

When obligations are implicit and enforcement is discretionary, failure spreads through blame, litigation, and power struggles. Preventing failure becomes a survival strategy, even at the cost of progress.

Hyperspace inverts this incentive.

By bounding failure mechanically, it removes the need to fear it.

MECHANISM: FAILURE AS STABILITY

At the mechanical level, failure stabilizes coordination through three properties:

1. **Deterministic transition** — failure occurs without interpretation
2. **Scope containment** — only dependent states are affected
3. **Immediate release** — blocked paths clear automatically

These properties prevent failure from becoming political.

FROM FAILURE TO FORKING

When failure repeats, persistence becomes irrational.

At that point, the correct response is not escalation, but divergence.

The next chapter examines forking, not as schism or betrayal, but as the system-level response to persistent infeasibility.

CHAPTER 20
FORKS AS RESOLUTION, NOT SCHISM

In belief-based systems, divergence is treated as failure.

When participants disagree, leadership intervenes. Compromise is sought. Authority asserts itself to preserve unity. Forking is framed as betrayal, fragmentation, or loss of legitimacy.

This framing is backwards.

Forks are not breakdowns in coordination.

They are its resolution.

WHY DIVERGENCE IS INEVITABLE

No coordination system can accommodate all objectives indefinitely.

Constraints change. Resources shift. Participants discover incompatibilities that were not visible at specification time. What once appeared coherent becomes infeasible.

In discretionary systems, these incompatibilities are negotiated.

Negotiation does not resolve divergence.

It suppresses it.

Suppressed divergence accumulates pressure until it erupts catastrophically.

FORKING AS STRUCTURAL SEPARATION

A fork is not an argument.

It is a separation of state spaces.

When an execution graph no longer admits a viable path for all participants, the correct response is not arbitration, prioritization, or compromise. It is duplication and divergence.

Each fork defines its own constraints.

Coordination continues independently.

NO PERMISSION REQUIRED

Forking in Hyperspace does not require approval.

There is no authority empowered to deny it. There is no legitimacy process to satisfy. Forking occurs when participants choose to instantiate a new execution geometry.

This choice is explicit.

It carries cost.

But it is final.

THE COST SIGNAL OF FORKING

Forking is intentionally expensive.

It requires re-specification of constraints, re-acquisition of commitments, and re-establishment of execution paths. These costs prevent frivolous divergence.

Forks occur only when coordination within the existing geometry has become irrational.

This is a feature.

Cheap forks produce fragmentation.

Expensive forks produce clarity.

FORKS VS. ESCALATION

Traditional systems respond to divergence through escalation.

More meetings. Higher-level review. Emergency authority. Temporary overrides. Each step concentrates power and erodes trust.

Forking bypasses escalation entirely.

Instead of deciding who wins, the system allows incompatible objectives to separate.

No one loses.

Coordination resumes.

PRESERVING FINALITY

Forking does not invalidate prior execution.

Completed transitions remain final. Commitments already resolved do not reopen. History is not rewritten.

Each fork inherits the past.

Only the future diverges.

This preserves trust in finality while permitting adaptation.

WHEN FORKS SHOULD NOT OCCUR

Forking is not a substitute for execution.

It should not be used to escape difficult work, evade deadlines, or avoid failure. Because forks impose cost, using them defensively is irrational.

Forks are appropriate only when constraints themselves are incompatible.

Misuse is self-limiting.

MECHANISM: FORKING AS COORDINATION

At the mechanical level, forks stabilize coordination through three properties:

1. **State space duplication** — divergence without interference
2. **Costly initiation** — prevents trivial fragmentation
3. **Inherited finality** — preserves trust in prior execution

Together, these properties replace political schism with mechanical separation.

FROM DIVERGENCE TO TERMINATION

Forking resolves incompatibility by separation.

Sometimes, separation is unnecessary.

Sometimes, coordination itself should end.

The next chapter examines when execution graphs should be allowed to die, and why termination is often the most honest outcome.

CHAPTER 21 WHEN COORDINATION SHOULD DIE

Persistence is often mistaken for virtue.

Systems celebrate endurance. Projects are praised for surviving setbacks. Organizations valorize resilience without examining whether continued coordination remains rational.

This instinct is destructive.

Not all coordination should continue.

Some coordination should end.

THE MYTH OF SALVAGEABILITY

Belief-based systems assume that coordination can always be repaired.

If progress stalls, more alignment is sought. If execution fails, governance intervenes. If participants disengage, incentives are adjusted. Termination is treated as defeat.

This assumption is false.

Some execution graphs admit no viable path forward under

existing constraints. No amount of negotiation can create feasibility where none exists.

Persistence becomes denial.

SUNK COST AS A COORDINATION TRAP

Once effort has been invested, systems resist termination.

Past work becomes leverage. Participants argue that abandoning coordination would waste prior investment. This logic is emotionally compelling but mechanically irrelevant.

Execution does not care about history.

Only future feasibility matters.

When coordination is sustained solely to justify past cost, the system consumes additional resources without increasing the probability of success.

This is collapse in slow motion.

TERMINATION AS A FIRST-CLASS TRANSITION

Hyperspace encodes termination explicitly.

An execution graph may reach a state where no remaining transitions are viable. When this occurs, the system transitions deterministically into termination.

Termination is not failure.

It is resolution.

No further commitments may be acquired. No additional execution occurs. The system closes.

DISTINGUISHING FAILURE, FORK AND DEATH

Hyperspace separates three outcomes that traditional systems conflate:

- **Failure** — a single path proves infeasible
- **Fork** — incompatible objectives require divergence
- **Termination** — no feasible future exists

Each outcome has a distinct response.

Confusing them produces pathology.

WHY TERMINATION IS RARE BUT NECESSARY

Well-specified systems rarely terminate.

Most failures resolve locally. Forks handle divergence. Termination occurs only when constraints were misdesigned or external conditions have invalidated the entire geometry.

When it does occur, delaying termination compounds loss.

Ending coordination early preserves optionality elsewhere.

NO RESURRECTION

Once terminated, a coordination instance cannot be revived.

Resurrection would require reinterpretation of constraints, renegotiation of commitments, or reversal of finality. Each would reintroduce governance.

If coordination must continue, it must restart.

Cleanly.

This restart is not continuity.

It is replacement.

TERMINATION WITHOUT BLAME

Termination does not assign fault.

Participants may have acted competently. Constraints may have been reasonable at specification time. External conditions may have changed.

The system does not inquire.

It records that coordination is no longer feasible.

This clarity prevents narrative conflict.

MECHANISM: TERMINATION AS COORDINATION HYGIENE

At the mechanical level, termination stabilizes the ecosystem through three properties:

1. **Explicit end states** — no indefinite execution
2. **Automatic closure** — no discretionary prolongation
3. **Clean restart requirement** — prevents zombie coordination

Together, these properties prevent collapse from spreading.

FROM TERMINATION TO TRADEOFF

Termination accepts loss.

This acceptance has cost.

The next chapter examines the deepest cost of Hyperspace: irreversibility, and why a system that refuses rollback must also refuse consolation.

CHAPTER 22
THE COST OF IRREVERSIBILITY

Irreversibility is the most demanding property of Hyperspace.

It is also the most easily misunderstood.

To remove governance from execution, outcomes must be final. To enforce constraint mechanically, transitions must not be revisited. To eliminate arbitration, there must be nothing left to appeal.

This has cost.

Hyperspace does not deny it.

WHAT IRREVERSIBILITY REMOVES

Irreversibility removes several features that belief-based systems rely on:

- Mercy after failure
- Rollback after error
- Narrative repair after conflict
- Contextual reconsideration

These features feel humane.

They are also discretionary.

Once reintroduced, they require judgment, authority, and exception handling. Execution becomes negotiable again.

Irreversibility is the price of coherence.

WHO THIS SYSTEM IS NOT FOR

Hyperspace is not designed for every form of collaboration.

It is inappropriate where:

- Outcomes are primarily expressive or relational
- Learning requires frequent reinterpretation of success
- Participants cannot accept definitive loss
- Ethical judgment must intervene during execution

In these domains, governance is not a flaw.

It is the objective.

Hyperspace should not be applied.

ERROR IS PAID IN FULL

In irreversible systems, mistakes are expensive.

A mis-specified constraint does not soften with empathy. A poorly chosen deadline does not adjust to effort. A bad commitment resolves exactly as encoded.

This severity disciplines design.

Participants learn to specify carefully, commit conservatively, and accept consequence fully.

The system does not forgive.

It clarifies.

WHY ROLLBACK DESTROYS TRUST

Rollback appears safe.

Mechanically, it is corrosive.

If outcomes can be undone, participants hedge. They delay dependent work. They seek informal guarantees. Finality erodes.

Rollback converts execution into provisional theater.

Irreversibility converts it into reality.

IRREVERSIBILITY AND HUMAN COST

The absence of mercy is uncomfortable.

Systems that permit appeal allow participants to preserve dignity after failure. Hyperspace does not offer this consolation.

Instead, it offers clarity.

Participants are not judged, but they are also not protected.

This tradeoff is intentional.

THE LIMIT OF MECHANICAL SYSTEMS

Irreversibility marks the boundary beyond which mechanical coordination should not extend.

When outcomes must be revisited for moral, emotional, or contextual reasons, constraint must yield to governance.

Hyperspace does not claim universality.

It claims precision.

MECHANISM: WHY IRREVERSIBILITY IS NON-NEGOTIABLE

At the mechanical level, irreversibility sustains coordination through three properties:

1. **Trust in finality** — participants can act without hedging
2. **Elimination of appeal** — no authority re-enters execution
3. **Design discipline** — constraints are specified with care

Removing any one reintroduces governance.

ACCEPTING THE COST

Hyperspace asks participants to accept loss without consolation.

In exchange, it offers something rarer:

- Execution without politics.
- Coordination without belief.
- Failure without blame.

These are not free.

They require irreversibility.

CLOSING PART IV

With irreversibility acknowledged, Part IV concludes.

The system has shown how it fails, forks, terminates, and refuses mercy.

What remains is not reassurance.

It is commitment.

The next part examines what it takes to adopt such a system, and why most attempts will fail before they begin.

PART VI

ADOPTION, LIMITS & REALITY

Up to this point, this book has treated Hyperspace as a mechanical system.

It has specified how coordination can occur without trust, governance, or discretionary authority. It has shown how execution resolves, how failure propagates, how divergence forks, and how coordination terminates without appeal. Taken together, these chapters describe a system that is internally coherent and mechanically complete.

That is not enough.

A system that works in theory but fails in adoption is not merely incomplete, it is misleading. The purpose of this part is therefore not to extend the thesis, but to constrain it.

Part VI examines what happens when Hyperspace is introduced into real organizations, real social structures, and real power dynamics. It does not ask how to convince institutions to adopt it. It asks why most cannot, what is lost by those that do, and how misuse reliably destroys the very properties that make the system work.

This is not a persuasive section.

It is a sorting mechanism.

Readers who arrive here expecting a rollout strategy, a change-management framework, or a gentler version of the system should stop. Hyperspace does not become safer when softened. It becomes incoherent.

The chapters that follow describe the narrow conditions under which adoption is possible, the social costs that must be paid up front, the failure modes of compromise, and the evidence that would falsify the thesis entirely.

If the system survives these constraints, it may be worth building.

If it does not, it should be abandoned.

This is not where the system expands, but where it sheds.

CHAPTER 23

WHY MOST ORGANIZATIONS CANNOT USE HYPERSPACE

Hyperspace is not difficult to understand.

It is difficult to survive.

The barrier to adoption is not technical sophistication, tooling maturity, or computational scale. The barrier is organizational identity.

Most organizations cannot use Hyperspace because doing so would require them to stop being what they are.

COORDINATION IS NOT MOST ORGANIZATIONS' PRIMARY GOAL

Organizations rarely exist to optimize execution.

They exist to:

- Preserve leadership legitimacy
- Maintain internal harmony
- Allocate credit and blame
- Absorb uncertainty without exposing error
- Survive political pressure

These goals are orthogonal to mechanical coordination.

Hyperspace optimizes only one thing: **feasible execution under constraint**.

Any organization that requires ambiguity, discretion, or narrative control during execution will experience Hyperspace as hostile.

DISCRETION IS NOT A BUG – IT IS THE PRODUCT

In most institutions, discretion is not accidental.

Managers are valued precisely because they can reinterpret rules, override procedures, and absorb conflict without finality. Executives exist to make exceptions when systems fail.

Hyperspace eliminates this role.

There are no overrides. No emergency powers. No contextual mercy.

For organizations built around discretionary authority, adoption is self-erasure.

LEGITIMACY DEPENDS ON EXPLANATION

Traditional organizations derive legitimacy from explanation.

When outcomes disappoint, leaders explain why the result was still correct, unavoidable, or morally justified. This narrative repair preserves trust.

Hyperspace produces outcomes without explanation.

Results occur because constraints resolved. Failure occurs because paths were infeasible. There is nothing to reinterpret.

Organizations that require narrative legitimacy will find this intolerable.

STATUS SYSTEMS CANNOT SURVIVE MECHANICAL FINALITY

Most organizations are status machines.

Titles, seniority, influence, and reputation mediate who decides, who is protected, and whose failure is excusable.

Hyperspace ignores status entirely.

Constraints apply identically. Commitments resolve identically. Failure is public and final.

This collapses informal hierarchies.

Those hierarchies will fight back.

ERROR EXPOSURE IS POLITICALLY FATAL

Hyperspace exposes mis-specification immediately.

Bad constraints fail fast. Poor commitments resolve visibly. There is no buffer period for explanation or repositioning.

In political organizations, this is dangerous.

Leaders depend on time to manage perception. Hyperspace removes that time.

As a result, adoption attempts are quietly sabotaged long before technical failure occurs.

SCALE MAGNIFIES INCOMPATIBILITY

Small teams can sometimes tolerate Hyperspace-like execution informally.

At scale, incompatibilities become explicit.

- Legal departments demand exceptions

- HR demands compassion
- Compliance demands overrides
- Leadership demands discretion

Each demand reintroduces governance.

The system collapses not because it fails, but because it works.

MISREADING THIS AS A CRITIQUE

This chapter is not an indictment of organizations.

Most institutions are not broken. They are optimized for survival, legitimacy, and social cohesion.

Hyperspace optimizes for none of these.

It optimizes for execution.

The incompatibility is structural, not moral.

WHO CAN USE HYPERSPACE

The set of viable adopters is narrow:

- Systems where execution dominates identity
- Environments where failure is acceptable and informative
- Domains where authority cannot be trusted
- Coordination problems with adversarial pressure

Outside these conditions, Hyperspace is unnecessary or harmful.

READER SORTING

If this chapter feels uncomfortable, that is intentional.

Hyperspace is not meant to be adopted gradually, politely, or universally.

Most readers should stop here.

Those who do not already know why they need this system will attempt to soften it.

That attempt will fail.

FROM INCOMPATIBILITY TO COST

If incompatibility were the only barrier, adoption would be simple.

It is not.

Even organizations that could use Hyperspace must pay a significant social cost.

The next chapter examines that cost directly.

CHAPTER 24
THE SOCIAL COST OF ADOPTION

Technical systems are rarely rejected for technical reasons.

They are rejected because they threaten existing social equilibria.

Hyperspace is no exception.

Adopting it exacts a cost that cannot be amortized, delegated, or deferred.

It is paid socially.

POWER IS LOST BEFORE EFFICIENCY IS GAINED

Hyperspace does not remove inefficiency first.

It removes discretion.

This means that the earliest adopters do not experience productivity gains. They experience loss of informal authority. Decisions that once relied on judgment, negotiation, or seniority become mechanically constrained.

Those who previously mediated conflict become redundant.

This loss precedes any visible benefit.

Most organizations abandon adoption at this stage.

THE END OF HEROIC INTERVENTION

Many organizations are stabilized by heroism.

When systems fail, experienced actors intervene, override process, and rescue outcomes. These interventions generate loyalty and status.

Hyperspace eliminates rescue.

Failures resolve automatically. Missed deadlines expire. Infeasible paths terminate. There is no moment for heroism.

This is experienced as cruelty.

NO PLACE TO HIDE AFTER FAILURE

In discretionary systems, failure is contextualized.

Effort is emphasized. Intent is weighed. External factors are cited. Responsibility is diffused.

Hyperspace offers no such insulation.

Failure is recorded as infeasibility.

Participants are not blamed, but they are also not protected.

For many, this exposure is intolerable.

SOCIAL TRUST VS. MECHANICAL TRUST

Organizations often conflate trust with familiarity.

They trust people because they know them, have worked with them, or share values. This trust allows flexibility and forgiveness.

Hyperspace replaces this with mechanical trust.

Outcomes are trusted not because participants are virtuous, but because constraints enforce resolution.

This shift feels alien.

It is experienced as dehumanization, even when outcomes improve.

LEADERSHIP WITHOUT EXCEPTION

Leadership in Hyperspace is constrained leadership.

Leaders specify constraints, commit resources, and accept outcomes. They cannot intervene during execution.

This demands a form of accountability most leaders have never practiced.

Adoption therefore requires leadership willing to lose face publicly.

Such leaders are rare.

ORGANIZATIONAL GRIEF

When Hyperspace is adopted seriously, something is lost.

Roles disappear. Informal influence evaporates. Longstanding negotiation rituals become irrelevant.

This produces grief.

Organizations that do not acknowledge this loss attempt to compensate with culture, language, or symbolism.

These compensations fail.

The loss must be accepted, not repaired.

WHY THIS COST CANNOT BE OFFSET

The social cost of adoption cannot be offset by incentives.

No amount of compensation restores discretion. No bonus replaces authority. No cultural program recreates exception handling.

Attempts to soften the transition reintroduce governance.

The system collapses quietly.

ADOPTION AS SELECTION

Successful adoption looks less like rollout and more like attrition.

Participants who cannot tolerate finality leave. Leaders who require discretion exit. Remaining actors reorganize around constraint.

This is not failure.

It is alignment.

FROM COST TO MISUSE

Even organizations willing to pay this cost often attempt to hedge.

They adopt partially.

They preserve exceptions.

They pilot without commitment.

The next chapter explains why these attempts fail, and why partial adoption is worse than none at all.

CHAPTER 25
WHY PARTIAL ADOPTION FAILS

Partial adoption is the most common response to threatening systems.

It is also the most destructive.

Organizations that attempt to adopt Hyperspace selectively do not mitigate risk.

They guarantee collapse.

CONSTRAINT CANNOT COEXIST WITH DISCRETION

Hyperspace replaces judgment with constraint.

Partial adoption preserves judgment in the name of safety, culture, or transition. This creates a dual system:

- Mechanical execution for some paths
- Discretionary override for others

The result is not balance.

It is incoherence.

Participants learn immediately which outcomes are final and which are negotiable. Trust migrates to discretion. Constraints are gamed.

Execution becomes political again.

HYBRID SYSTEMS REINTRODUCE ARBITRATION

Any exception requires an arbiter.

Partial adoption therefore recreates governance automatically. Someone must decide when the system applies and when it does not.

This authority is informal at first.

It becomes institutionalized quickly.

At that point, Hyperspace no longer governs execution. It decorates it.

PILOTS FAILS BECAUSE THEY CANNOT FAIL

Organizations often propose pilots.

Pilots are protected environments. Failure is expected, contained, and explained away. This directly contradicts the system's core requirement: real consequence.

A pilot that cannot terminate, fork, or impose irreversible loss teaches participants that outcomes are provisional.

When scaled, the system is not believed.

SELECTIVE FINALITY DESTROYS TRUST

Finality must be universal.

If some commitments resolve mechanically while others are revisited, participants hedge everywhere. Downstream work stalls. Dependencies are delayed.

Selective irreversibility produces the worst of both worlds:

- The rigidity of constraint
- The politics of discretion

Trust collapses faster than under either system alone.

MANAGERS WILL REINSERT EXCEPTIONS

This is not a moral failure.

Managers are rewarded for preventing visible loss. Under pressure, they will intervene to protect people, outcomes, or reputation.

Each intervention teaches the organization that the system is optional.

The more pressure increases, the faster the system unravels.

PARTIAL ADOPTION AS SELF-DECEPTION

Organizations tell themselves they are "learning" the system.

In reality, they are avoiding commitment.

Hyperspace cannot be learned safely. It can only be learned honestly.

Anything else is rehearsal without consequence.

WHY PARTIAL ADOPTION IS WORSE THAN NONE

A failed partial adoption discredits the system.

Participants conclude that mechanical coordination is brittle, inhumane, or impractical. These conclusions are drawn not from the system itself, but from its sabotage.

This prevents future honest adoption.

The damage persists.

THE ONLY VIABLE ADOPTION MODE

Hyperspace admits exactly one adoption mode:

- Full constraint
- Universal finality
- No exception paths

Anything less is not adoption.

It is simulation.

FROM FAILURE TO FALSIFICATION

If partial adoption fails, the failure is meaningless.

If full adoption fails, the thesis is threatened.

The final chapter defines what such a failure would look like, and what evidence would falsify the system entirely.

CHAPTER 26
WHAT WOULD FALSIFY THE THESIS

A manifesto that cannot be falsified is not a system.

It is a belief.

Hyperspace does not claim inevitability. It claims necessity under specific conditions. If those conditions can be met by other means, Hyperspace is redundant.

This chapter defines what evidence would invalidate the thesis entirely.

THE CORE CLAIM

The central claim of Hyperspace is narrow:

Large-scale coordination under adversarial conditions cannot rely on discretion, consensus, or governance during execution.

Everything else follows from this.

To falsify Hyperspace, one must falsify this claim.

VALID FALSIFIERS

Any of the following would invalidate the need for Hyperspace:

1. Scalable Discretion

Demonstrate a system that:

- Operates at scale
- Permits discretionary intervention during execution
- Does not degrade into political arbitration
- Maintains predictable outcomes under pressure

Such a system would eliminate the need for mechanical finality.

2. Consensus Without Latency

Provide evidence that:

- Consensus mechanisms can converge rapidly
- Alignment persists under adversarial incentives
- Disagreement does not accumulate hidden pressure

If consensus can scale without delay or collapse, constraint is unnecessary.

3. Neutral Governance at Execution time

Identify a governance structure that:

- Intervenes during execution
- Remains neutral across outcomes
- Does not concentrate power

- Does not erode trust

This would contradict the claim that governance and execution are incompatible.

4. Reversibility Without Hedging

Demonstrate a system where:

- Outcomes can be rolled back
- Participants do not hedge or delay dependent work
- Trust in finality is preserved

If rollback does not corrupt execution, irreversibility is unnecessary.

INVALID COUNTEREXAMPLES

Certain arguments do not falsify Hyperspace:

- Small teams with high trust
- Short-lived projects
- Systems protected by monopoly power
- Environments without adversarial pressure

These cases fall outside the problem domain.

They are not counterexamples.

WHAT FAILURE WOULD LOOK LIKE

If Hyperspace were adopted fully and honestly, failure would look specific:

- Constraints encode correctly but execution still degrades

- Finality is respected but coordination stalls
- Forking occurs yet outcomes worsen
- Termination triggers despite feasible paths

Such evidence would require revision or abandonment of the thesis.

WHY THIS HAS NOT HAPPENED YET

The absence of falsification is not proof.

It reflects the rarity of systems that:

- Remove governance entirely
- Accept irreversibility
- Expose error immediately

Most attempts fail earlier, for social reasons already described.

OPEN TO REFUTATION

Hyperspace invites refutation.

If a better system exists, it should replace this one.

Progress does not require loyalty.

It requires evidence.

CLOSING PART VI

This part has not argued for adoption.

It has argued for restraint.

Most readers should walk away.

Those who remain should do so without illusion.

FROM LIMITS TO IMPLEMENTATION

With boundaries defined, the manifesto is complete.

What remains is optional: translating thesis into platform.

That translation does not strengthen the argument.

It merely tests it.

In the final part, these principles converge in a concrete system: Metaspace as a falsifiable instance of the Hyperspace thesis.

PART VII

FROM THESIS TO PLATFORM

Up to this point, the argument has been complete.

The Hyperspace Manifesto has identified the coordination problem, eliminated existing solutions, specified a mechanical alternative, defined its failure modes, constrained its adoption, and made its claims explicitly falsifiable. Nothing in the preceding chapters requires implementation in order to be internally coherent.

And yet, stopping here would be evasive.

A system that claims inevitability under certain conditions must be willing to encounter those conditions directly. Theory alone cannot absorb adversarial behavior, operational friction, or the cumulative weight of real consequences. For that, execution is required.

This final part does not expand the thesis.

It collapses it.

Part VII exists to describe what it means to instantiate Hyperspace without diluting it, what must be encoded, what must be refused, and where governance is allowed to exist only at the boundary. It is not an invitation to build enthusiastically, nor a promise that building will succeed.

It is a declaration of commitment.

The chapter that follows introduces Metaspace not as a product, platform, or future vision, but as a concrete test of the claims made in this book. If the system behaves as specified, the thesis stands provisionally. If it does not, the thesis fails.

Nothing beyond that outcome is guaranteed.

This is not where the argument becomes optimistic.

It is where it becomes accountable.

A manifesto that cannot be instantiated
is indistinguishable from theory.

CHAPTER 27
METASPACE

The Hyperspace Manifesto is not complete without execution.

Metaspace.app exists as a concrete instance of the principles described in this book. It is not a conclusion. It is a testbed.

Every claim made in the preceding chapters terminates in an implementation requirement. Every principle must survive contact with infrastructure. Anything that cannot be encoded is intentionally excluded.

This chapter grounds Hyperspace in reality, not as a product roadmap, but as a set of irreversible commitments embodied by *Metaspace*.

THE COST OF INSTANTIATION

Implementing a Hyperspace system requires abandoning several assumptions common to contemporary platforms:

- That users must be retained at all costs
- That ambiguity increases adoption
- That governance must be participatory
- That discretion is a feature

Metaspace rejects these assumptions deliberately.

It is not designed to be comfortable.

It is designed to be correct.

VEKTAGRAF AS STRUCTURAL SUBSTRATE

Vektagraf provides the structural foundation for Metaspace.

Projects, tasks, dependencies, and participants are represented as explicit graph objects. Relationships are directional and constrained. State transitions are recorded, not inferred.

This graph-first model is not an optimization.

It is a requirement.

Without an explicit graph, coordination collapses back into narrative and interpretation.

PRIVACY FIRST AS CONTROL LAYER

Privacy First supplies the enforcement boundary.

Identity, permissions, capability disclosure, and verification operate without exposing participants to unnecessary risk. Control is exercised through capability possession rather than identity assertion.

This enables a system that is both trustless and private, two properties often assumed to be in tension.

They are not.

NO ADMIN PANEL

Metaspace has no superuser.

There is no administrative override, no emergency backdoor, and no hidden escalation path. If intervention were possible, authority would merely be delayed, not removed.

The absence of an admin role is not ideological.

It is structural.

BUILDING REQUIRES COMMITMENT

To build Metaspace is to accept constraint.

Developers cannot patch around uncomfortable outcomes. Operators cannot intervene when progress slows. Designers cannot soften failure modes to protect sentiment.

The system must be allowed to behave exactly as specified, even when that behavior is inconvenient.

Anything less reintroduces discretion.

CONTRIBUTORS, NOT CUSTOMERS

Participants in Metaspace are contributors, not users.

They are not promised ease, safety, or fairness as feelings. They are promised deterministic outcomes under known constraints.

Those unwilling to accept mechanical resolution will self-select out. This is not a retention problem.

It is a filtering mechanism.

ADOPTION THROUGH NECESSITY

Metaspace is unlikely to spread through persuasion or marketing.

It will be adopted where existing coordination systems fail catastrophically: high-stakes collaboration among strangers, capital-intensive projects without trust, and long-horizon work where governance collapse is common.

Adoption will be driven by necessity, not enthusiasm.

EXPLICIT FAILURE MODES

Metaspace does not claim to eliminate failure.

It claims to make failure unambiguous.

Projects can stall. Tasks can expire. Forks can fragment effort. These outcomes are visible and final.

Opacity is removed.

WHAT HYPERSPACE DEMANDS

From builders, Hyperspace demands restraint.

From contributors, it demands precision.

From organizations, it demands the surrender of control.

These demands are non-negotiable.

A FALSIFIABLE CLAIM

Metaspace exists to test the claims made in *The Hyperspace Manifesto*.

If the platform functions as described, the thesis stands.

If it does not, the thesis fails.

There is no rhetorical escape.

CLOSING THE LOOP

This book began by arguing that trust, governance, and authority were compensations for insufficient systems.

It ends with an instance that removes them.

Hyperspace is not a promise of a better future.

Metaspace is a test.

If coordination can occur without trust, permission, or governance, then those concepts were scaffolding, not fundamentals.

What remains is execution.

That is the final claim of *The Hyperspace Manifesto*.

EPILOGUE

CONSTRAINT, NOT OPTIMISM

Manifestos are written when existing language fails.

They emerge not from optimism, but from constraint: from the repeated observation that prevailing systems cannot be corrected incrementally. They must be replaced.

Metaspace did not begin as a product idea. It emerged as an inevitability.

A LINEAGE OF CONSTRAINT

This work follows a clear lineage.

The Hyperstore Manifesto argued that commerce collapses when authority and inventory are abstracted away from execution. It proposed systems where value, availability, and fulfillment are structurally bound rather than contractually asserted.

The Hyperproof Manifesto extended that argument further: authority itself could no longer be entrusted to institutions or individuals. It had to be encoded.

Vektagraf and *Privacy First* were not experiments. They were **falsifiable implementations**.

Metaspace exists because those implementations held.

FROM AUTHORITY TO COORDINATION

If authority can be encoded, coordination becomes the next failure point.

Trustless commerce without trustless collaboration merely shifts risk. It does not eliminate it. Organizations still stall. Projects still collapse under governance. Human bottlenecks persist.

Metaspace addresses the final unresolved layer: how strangers act together without belief, alignment, or persuasion.

WHY THIS IS NOT A MOVEMENT

Movements require agreement.

Metaspace does not.

There is nothing to rally around and no values to adopt. Participation requires only acceptance of constraints. Those constraints are explicit and enforced.

If the system produces outcomes, it will be used. If it does not, it will be abandoned.

This is not ideology.

It is selection.

THE DISAPPEARANCE OF THE BUILDER

The ultimate success of *Metaspace* would render its creators irrelevant.

There are no stewards to trust, no roadmap to follow, no leadership to defer to. The system does not improve because someone wills it to. It improves only when change executes.

This is not decentralization as branding.

It is abdication.

WHAT COMES AFTER TRUST

Trust was never a virtue. It was a workaround.

It emerged to compensate for systems that could not enforce their own rules. As enforcement improves, trust becomes unnecessary, and eventually dangerous.

Metaspace is not anti-human. It is anti-ambiguity.

It does not remove agency. It removes leverage.

A NARROW AUDIENCE

This book is not written for everyone.

It is written for those who have already attempted collaboration at scale and found that culture, incentives, and governance fail under pressure.

It is written for builders who are willing to give up control in exchange for determinism.

Most will not.

That is expected.

THE ONLY REMAINING QUESTION

The question *Metaspace* poses is not whether this system is desirable.

It is whether it is possible.

That question cannot be answered by argument.

It can only be answered by execution.

CLOSING

If *Metaspace* works, then trust, governance, and authority were scaffolding.

If it fails, then they were fundamentals.

There is no third option.

The platform will decide.

That is sufficient.

ABOUT THE AUTHOR

Dom Jocubeit is a systems architect, theorist, and engineer whose work focuses on the mechanical enforcement of authority, value, and coordination in digital environments.

He is focused on the limits of large-scale coordination and is the author of *The Hyperstore Manifesto* and *The Hyperproof Manifesto*, both of which articulated theses that were later instantiated as functioning systems: Vektagraf and Privacy First. These platforms served not as demonstrations of ideology, but as falsifiable implementations of specific claims about markets, authority, and execution.

Dom is the creator of the **Hypermodern Theorem**, which formalizes the claim that trust, governance, and discretionary authority do not scale under adversarial conditions, and that sufficiently complex coordination systems must replace belief with constraint. The theorem provides the conceptual foundation for the work presented in *The Hyperspace Manifesto*, a coordination model based on explicit state geometry, irreversible commitments, and mechanical finality; and **Metaspace**, its concrete implementation and falsifiable testbed.

Dom's work sits at the intersection of distributed systems, economics, and institutional design. His research is concerned less with optimization and more with elimination: removing categories of failure by changing the structure of execution itself.

He writes for builders and system designers working in environments where trust is insufficient, authority is contested, and failure is costly. His work is intentionally narrow, unsentimental, and grounded in execution rather than persuasion.

linkedin.com/in/jocubeit
x.com/jocubeit

GLOSSARY

HYPERSPACE TERMS

This glossary defines terms as they are used *specifically within The Hyperspace Manifesto*. Where common words are repurposed, these definitions supersede conventional usage.

Arbitration
Any discretionary process used to resolve outcomes during execution. Arbitration requires judgment, authority, and exception handling. Hyperspace eliminates arbitration entirely.

Authority
The ability to intervene in execution based on status, role, or discretion. Authority is treated as a compensatory mechanism for insufficient systems.

Capability
A cryptographic or structural right to act within the system. Capabilities replace identity- or role-based permissioning.

Commitment
A mechanically binding declaration that locks state and resources until resolution. Commitments are not promises; they are state constraints.

Constraint
A non-negotiable rule encoded into the execution system that limits possible transitions. Constraint replaces consensus as the basis of coordination.

Coordination
The alignment of multiple participants toward a shared outcome. In Hyperspace, coordination is achieved mechanically, not socially.

Discretion
The ability to reinterpret rules, override outcomes, or apply contextual judgment during execution. Discretion is incompatible with mechanical finality.

Execution
The resolution of commitments through constrained state transitions. Execution is distinct from planning, governance, or interpretation.

Execution Graph
An explicit representation of tasks, dependencies, participants, and constraints as a directed graph. All coordination occurs through this structure.

Failure
A deterministic system state in which a transition cannot occur within defined constraints. Failure is explicit, local, and final.

Finality
The property that resolved outcomes cannot be revisited, appealed, or undone. Finality is required to eliminate governance during execution.

Fork
The duplication and divergence of an execution graph in response to incompatible constraints. Forking resolves divergence without arbitration.

Governance
Any system of decision-making that relies on legitimacy, voting, hierarchy, or consensus. Governance is permitted only outside execution.

Hybrid System
A coordination system that mixes mechanical constraint with discretionary override. Hybrid systems inevitably collapse back into governance.

Irreversibility
The inability to roll back outcomes after execution. Irreversibility disciplines design and preserves trust in finality.

Legitimacy
Perceived justification for authority or outcomes. Hyperspace does not rely on legitimacy; it relies on correctness.

Mechanical Trust
Trust derived from enforced constraints and deterministic resolution rather than belief in participant intent.

Metaspace (metaspace.app)
A concrete implementation of the Hyperspace thesis. Metaspace functions as a falsifiable testbed, not a product claim.

Narrative Repair
Post-hoc explanation used to preserve legitimacy after failure or conflict. Narrative repair is excluded from Hyperspace systems.

Partial Adoption
The selective application of constraint or finality while preserving discretion elsewhere. Partial adoption guarantees system failure.

Platform
An implementation substrate that encodes execution rules and constraints. Platforms test theses; they do not validate them rhetorically.

Privacy First (privacyfirst.id)
A control model in which identity, permissions, and verification are minimized and protected. Privacy First enables trustless coordination without exposure.

Reputation
A record of resolved execution outcomes, not a social status. Reputation emerges mechanically and carries no authority.

Status
Social ranking that influences discretion, protection, or credibility. Status is ignored by Hyperspace systems.

Termination
A deterministic end state in which no viable future transitions exist. Termination resolves coordination without blame.

Trust
Reliance on belief in intent, competence, or goodwill. Hyperspace minimizes trust requirements through constraint.

Vektagraf (vektagraf.com)
The graph-capable structural substrate underlying Metaspace. Vektagraf encodes projects, tasks, dependencies, and participants as explicit state.

This glossary is normative. Terms are not descriptive conveniences; they define system behavior. Misusing them reintroduces ambiguity.

Made in the USA
Monee, IL
04 November 2026

67391576R00125